"A TAUT, RIVETING NARRATIVE."
—*Detroit News & Free Press*

"Preston uses his considerable storytelling skills to show us the heroes who fought smallpox, not for money or glory but simply because they wanted to leave behind a better world than they had found."
—*Dallas Morning News*

"Vivid testimony . . . The alarms he raises are real ones. . . . With his genius for vivid detail and telling anecdote, Preston adds frissons of his own. . . . His real métier lies in intimate and exhaustive interviews with experts on the front line."
—*Newsday*

"Riveting . . . Better-than-fiction characters . . . Preston had terrific access to people and the facilities typically off-limits."
—*Atlanta Journal-Constitution*

"Gripping. Preston humanizes his science reportage by focusing on individuals—scientists, physicians, government figures. That, and a flair for teasing out without overstatement the drama in his inherently compelling topics, plus a prose style that's simple and forceful, make this book as exciting as the best thriller."
—*Publishers Weekly* (starred review)

Please turn the page for more reviews. . . .

**"PRESTON CAPTIVATES . . .
A frighteningly real account of the virus
and its potential to explode globally."
—*Cleveland Plain Dealer***

"Compelling . . . Preston charts the tragic miscalculations and the geopolitical maneuvers that led from the triumph of eradication to the possible threat of a deliberate epidemic. . . . Preston is a master at explaining what's important. . . . A reminder of the lifesaving promise of global cooperation."
—*San Jose Mercury News*

"Darkly entertaining . . . Fascinating, frightening, and important. It reads like a thriller, but the demons are real. . . . Read this book and pray that its heroes can lock the demon back in the freezer."
—JONATHAN WEINER
Author of *The Beak of the Finch*

"Entertaining characters . . . Impressive journalistic research . . . Filled with smallpox information you didn't get from the news . . . Preston gives the story amazing new depth and tension."
—*Chicago Tribune*

"LYRICAL AS WELL AS EXPLANATORY . . .

Preston is a helpful guide, translating complex scientific situations into everyday language."
—*St. Louis Post-Dispatch*

"[Blends] terror, technology, and trivia . . . [Preston] has probably done more than any other writer to establish a nationwide imperative to think about infectious agents as global threats and potential weapons."
—*The New York Times Book Review*

"Riveting . . . Startling new insights into the government's reaction to the anthrax mailings."
—*Hartford Courant*

"Richard Preston has brought us another book that reads like a top-notch thriller. Would that it were fiction.'
—LAURIE GARRETT
Author of *The Coming Plague*

"The bard of biological weapons captures the drama of the front lines."
—RICHARD DANZIG
Former secretary of the navy

Also by Richard Preston

FIRST LIGHT
AMERICAN STEEL
THE HOT ZONE
THE COBRA EVENT
THE BOAT OF DREAMS
THE WILD TREES

THE
DEMON
IN THE
FREEZER

A TRUE STORY

Richard Preston

BALLANTINE BOOKS • NEW YORK

The author expresses his gratitude to the Alfred P. Sloan Foundation for a research grant that helped in the completion of this book.

A Ballantine Book
Published by The Random House Publishing Group
Copyright © 2002 by Richard Preston

Published in the United States by Ballantine Books, an imprint of The Random House Publishing Group, a division of Random House, Inc., New York, and simultaneously in Canada by Random House of Canada Limited, Toronto.

Ballantine and colophon are registered trademarks of Random House, Inc.

Portions of this book appeared in different form in *The New Yorker*.

www.ballantinebooks.com

ISBN 0-345-46663-2

Manufactured in the United States of America

First Random House Edition: October 2002
First Ballantine Mass Market Edition: September 2003

OPM 9 8 7

This book is lovingly dedicated to Michelle

Chance favors the prepared mind.
—LOUIS PASTEUR

Contents

SOMETHING IN THE AIR   1

THE DREAMING DEMON   27

TO BHOLA ISLAND   61

THE OTHER SIDE OF THE MOON   99

A WOMAN WITH A PEACEFUL LIFE   135

THE DEMON'S EYES   177

THE ANTHRAX SKULLS   197

SUPERPOX   261

Glossary   285
Acknowledgments   291

SOMETHING
IN THE AIR

Journey Inward

IN THE EARLY NINETEEN SEVENTIES, a British photo retoucher named Robert Stevens arrived in south Florida to take a job at the *National Enquirer,* which is published in Palm Beach County. At the time, photo retouchers for supermarket tabloids used an airbrush (nowadays they use computers) to clarify news photographs of world leaders shaking hands with aliens or to give more punch to pictures of six-month-old babies who weigh three hundred pounds. Stevens was reputed to be one of the best photo retouchers in the business. The *Enquirer* was moving away from stories like "I Ate My Mother-in-Law's Head," and the editors recruited him to bring some class to the paper. They offered him much more than he made working for tabloids in Britain.

Stevens was in his early thirties when he moved to Florida. He bought a red Chevy pickup truck, and he put a CB radio in it and pasted an American-flag decal in the back window and installed a gun rack next to the flag. He didn't own a gun: the gun rack was for his fishing rods. Stevens spent a lot of time at lakes and canals around south Florida, where he would spin-cast for bass and panfish. He often stopped to drop a line in the water on his way to and from work. He became an American citizen. He would drink a Guinness or two in bars with his friends and explain the Constitution to them. "Bobby was the only English redneck I ever knew," Tom Wilbur, one of his best friends, said to me.

Stevens's best work tended to get the *Enquirer* sued. When the TV star Freddie Prinze shot himself to death, Stevens joined two photographs into a seamless image of Prinze and Raquel Welch at a party together. The implication was that they had been lovers, and this sparked a lawsuit. He enhanced a photograph of a woman with a long neck: "Giraffe Woman." Giraffe Woman sued. His most famous retouching job was on a photograph of Elvis lying dead in his coffin, which ran on the cover of the *Enquirer*. Elvis's bloated face looked a lot better in Stevens's version than it did in the handiwork of the mortician.

Robert Stevens was a kindhearted man. He filed the barbs off his fishing hooks so that he could release a lot of the fish he caught, and he took care of feral cats that lived in the swamps around his house. There was something boyish about him. Even when he was in his sixties, children in the neighborhood would knock on the door and ask his wife, Maureen, "Can Bobby come out and play?" Not long before he died, he began working for *The Sun,* a tabloid published by American Media, the company that also owns the *National Enquirer.* The two tabloids shared space in an office building in Boca Raton.

·

ON THURSDAY, September 27th, Robert Stevens and his wife drove to Charlotte, North Carolina, to visit their daughter Casey. They hiked at Chimney Rock Park, where each autumn brings the spectacular sight of five hundred or more migrating hawks soaring in the air at once, and Maureen took a photograph of her husband with the mountains behind him. By Sunday, Stevens was not feeling well. They left for Florida Sunday night, and he got sick to his stomach during the drive home. On Monday, he began running a high fever and became incoherent. At two o'clock on Tuesday morning, Maureen took him to the emergency room of the John F. Kennedy Medical Center in Palm Beach County. A doctor there thought he might have meningitis. Five hours later, Stevens started having convulsions.

The doctors performed a spinal tap on him, and the fluid came out cloudy. Dr. Larry Bush, an infectious-disease specialist, looked at slides of the fluid and saw that it was full of rod-shaped bacteria with flat ends, a little like slender macaroni. The bacteria were colored blue with Gram stain—they were Gram-positive. Dr. Bush thought, *anthrax.* Anthrax, or *Bacillus anthracis,* is a single-celled bacterial micro-organism that forms spores, and it grows explosively in lymph and blood. By Thursday, October 4th, a state lab had confirmed the diagnosis. Stevens's symptoms were consistent with inhalation anthrax, which is caused when a person breathes in the spores. The disease is extremely rare. There had been only eighteen cases of inhalation anthrax in the past hundred years in the United States, and the last reported case had been twenty-three years earlier. The fact that anthrax popped into Dr. Bush's mind had not a little to do with recent news reports about two of the September 11th hijackers casing airports around south Florida and inquiring about renting crop-dusting aircraft. Anthrax could be distributed from a small airplane.

Stevens went into a coma, and at around four o'clock in the afternoon of Friday, October 5th, he suffered a fatal breathing arrest. Minutes later, one of his doctors made a telephone call to the Federal Centers for Disease Control and Prevention—the CDC—in Atlanta, and spoke with Dr. Sherif Zaki, the chief of infectious-diseases pathology.

Sherif Zaki inhabits a tiny office on the second floor of Building 1 at the CDC. The hallway is made of white cinder block, and the floor is linoleum. The buildings of the CDC sit jammed together and joined by walkways on a tight little campus in a green and hilly neighborhood in northeast Atlanta. Building 1 is a brick oblong with aluminum-framed windows. It was built in the nineteen fifties, and the windows look as if they haven't been cleaned since then.

Sherif Zaki is a shy, quiet man in his late forties, with a gentle demeanor, a slight stoop in his posture, a round face, and pale green eyes distinguished by dazzling pupils, which give him a piercing gaze. He speaks precisely, in a low voice. Zaki went out into the hallway, where his pathology group often gathered to talk about ongoing cases. "Mr. Stevens has passed away," he said.

"Who's going to do the post?" someone asked. A post is a postmortem exam, an autopsy.

Zaki and his team were going to do the post.

EARLY THE NEXT MORNING, on Saturday, October 6th, Sherif Zaki and his team of CDC pathologists arrived in West Palm Beach in a chartered jet, and a van took them to the Palm Beach County medical examiner's office, which takes up two modern, one-story buildings set under palm trees on a stretch of industrial land near the airport. They went straight to the autopsy suite, carrying bags of tools and gear. The autopsy suite is a large, open room in the center of one of the buildings. Two autopsies were in progress. Palm Beach medical examiners were bending over opened bodies on tables, and there was an odor of fecal matter in the air, which is the normal smell of an autopsy. The examiners stopped work when the CDC people entered.

"We're here to assist you," Zaki said in his quiet way.

The examiners were polite and helpful but did not make eye contact, and Zaki sensed that they were afraid. Stevens's body contained anthrax cells, although he had not been dead long enough for the cells to become large numbers of spores. In any case, any spores in his body were wet, and wet anthrax spores are nowhere near as dangerous as dry spores, which can float in the air like dandelion seeds, looking for fertile ground.

The CDC people opened a door in the morgue refrigerator and pulled out a tray. The body had been zipped up inside a Tyvek body bag. Without opening the bag, they lifted the body up by the shoulders and feet and placed it on a bare metal gurney. They rolled the gurney into a supply room and closed the door behind them. They would do the autopsy on the gurney in a closed room, to prevent the autopsy tables from being contaminated with spores.

The chief medical examiner of Palm Beach County, Dr. Lisa Flannagan, was going to do the primary incisions, while Zaki and his people would do the organ exams. Flannagan is a slender, self-assured woman, with a reputation as a top-notch examiner. Everybody gowned up, and they put on N-100 biohazard masks, clear plastic face shields, hair covers, rubber boots, and three layers of gloves. The middle glove was reinforced with Kevlar. Then they unzipped the bag.

The CDC team lifted the body up, gripping it beneath the shoulders and legs, and someone snatched the bag out from underneath it. They lowered the body back onto the bare metal deck of the gurney. Stevens had been a pleasant-looking man with a cheerful appearance. He was a bluish color now, and his eyes were half open.

Heraclitus said that when a man dies, a world passes away. The terribly human look on the face of the deceased man disturbed Sherif Zaki. It was so hard to picture this man in life and then to connect that picture with the body on the gurney. This was the toughest thing for a prosector, and you never got over it, really. Zaki did not want to connect the living man with the body. You had to put it aside, and you could not think about it. His duty now was to identify the exact type of disease that Stevens had, to learn if he had inhaled spores or perhaps had become infected some other way. This might help save lives. Yet cutting into an unfathomed body was difficult, and after a hard post, Sherif Zaki would not feel like himself for a week afterward. "It's not an uplifting process," Zaki said to me.

The team rolled Stevens onto his side and inspected his back under bright lights for signs of cutaneous anthrax—skin anthrax. They didn't find any, and they laid him back down.

Dr. Flannagan took up a scalpel and pressed the tip of the blade on the upper left part of the chest under the shoulder. She made a curving incision that went underneath the nipples, across the chest, and up to the opposite shoulder. Then, starting at the top of the sternum, she made a straight incision down to the solar plexus. This made a cut that looked like a Y, but with a curved top. She finished it with a short horizontal cut across the solar plexus. The opening incision looked rather like the profile of a wineglass.

Dr. Flannagan grasped the skin of the chest, and pulled it upward, peeling it off. She laid the blanket of skin around the neck. She pulled the skin away from the sides of the chest, revealing the ribs and sternum. She took up a pair of gardening shears and cut the ribs one by one, snipping them in a wide circle around the sternum. This was to free the chest plate, the front of the rib cage. When she had finished cutting the ribs, she pushed her fingertips underneath the chest plate and pried it upward, as if she were raising a lid from a box.

As Flannagan lifted the chest plate, a gush of bloody fluid poured out from under the ribs and ran down over the body and poured over the gurney and onto the floor.

The chest cavity was engorged with bloody liquid. No one in the room had ever done a post on a person who had died of anthrax. Zaki had studied photographs of autopsies that had been done on anthrax victims in the Soviet Union, in the spring of 1979, after a plume of finely ground anthrax dust had come out of a bioweapons manufacturing facility in Sverdlovsk (Yekaterinburg) and had killed at least sixty-six people downwind, but the photographs had not prepared him for the sight of the liquid that was pouring out of this man's chest. They were going to have quite a time cleaning up the room. The bloody liquid was saturated with anthrax cells, and the cells would quickly start turning into spores when they hit the air.

Dr. Flannagan stood back. It was the turn of the CDC team.

The CDC people wanted to look at the lymph nodes in the center of the chest. Working gently with his fingertips, Zaki separated the lungs and pulled them to either side, revealing the heart. The heart and lungs were drowned in red liquid. He couldn't see anything. Someone brought a ladle, and they started spooning the liquid from the chest. They poured it off into containers, and ultimately they had ladled out almost a gallon of it.

Zaki worked his way slowly down into the chest. Using a scalpel, he removed the heart and parts of the lungs, which revealed the lymph nodes of the chest, just below the fork of the bronchial tubes. The lymph nodes of a healthy person are pale nodules the size of peas. Stevens's lymph nodes were the size of plums, and they looked exactly like plums—they were large, shiny, and dark purple, verging on black. Zaki cut into a plum with his scalpel. It disintegrated at the touch of the blade, revealing a bloody interior, saturated with hemorrhage. This showed that the spores that had killed Stevens had gotten into his lungs through the air.

When they had finished the autopsy, the pathologists gathered up their tools and placed some of them inside the body cavity. The scalpels, the gardening shears, scissors, knives, the ladle—the prosection tools were now contaminated with anthrax. The team felt that the safest thing to do with them would be to destroy them. They packed the body cavity with absorbent batting, stuffing it in around the tools, and placed the body inside fresh double body bags. Then, using brushes and hand-pump sprayers filled with chemicals, they spent hours decontaminating the supply room, the bags, the gurney, the floor—everything that had come into contact with fluids from the autopsy. Robert Stevens was cremated. Sherif Zaki later recalled that when he was ladling the red liquid from Stevens's chest, the word *murder* never entered his mind.

THE DAY BEFORE Robert Stevens died, a CDC investigation team led by Dr. Bradley Perkins had arrived in Boca Raton and had begun tracing Stevens's movements over the previous few weeks. They wanted to find the source of his exposure to anthrax. They believed that it would have to be a single point in the environment, because anthrax does not spread from person to person. They split into three search groups. One group flew off to North Carolina and visited Chimney Rock while the other two went around Boca Raton. They all had terrorism on their minds, but Perkins wanted the team to make sure they didn't miss a dead cow with anthrax that might be lying next to one of Stevens's fishing spots.

Working the telephones, they called emergency rooms and labs, asking for any reports of unexplained respiratory illness or of organisms from a medical sample that might be anthrax. A seventy-three-year-old man named Ernesto Blanco turned up. Blanco, who was in Cedars Medical Center in Miami with a respiratory illness, happened to be the head of the mail room at the American Media building, where Robert Stevens worked. Doctors had taken a nasal swab from him, and the swab produced anthrax on a petri dish. Blanco and Stevens had not socialized with each other. The only place where their paths crossed was inside the American Media building.

The zone of the suspected point source shrank abruptly, and the CDC team went to the American Media building with swab kits. (A swab kit is a plastic test tube that holds a sterile medical swab, which looks somewhat like a Q-tip and has a thin wooden handle. You swab an area of interest, and then you push the swab into the test tube, snap off the wooden handle, cap the test tube, and label it. Later, the swab is brushed over the surface of a petri dish, and micro-organisms captured by the swab grow there, forming spots and colonies.) When they were running very short of swabs, Perkins and his people made a decision to test the mail bin for the photo department of *The Sun*.

The swab from the mail bin proved to be rich with spores of anthrax. It was brushed over a petri dish full of blood agar—sheep's blood in jelly—and by late in the afternoon of the day the autopsy took place, colonies and spots of anthrax cells were growing vigorously on the blood. The spots were pale gray, and they sparkled like pow-dered glass—they had the classic, glittery look of anthrax. Something full of spores must have arrived in the mail. It meant that the point source of the outbreak was nothing in nature. On Sunday night, October 7th, Brad Perkins tele-phoned the director of the CDC, Dr. Jeffrey Koplan. "We have evidence for an intentional cause of death of Robert Stevens," he said to Koplan. "The FBI needs to come into this full force."

Communiqué from Nowhere

OCTOBER 15, 2001

AT TEN O'CLOCK on a warm autumn morning in Washington, D.C., a woman—her name has not been made public—was opening mail in the Hart Senate Office Building, on Delaware Avenue. She worked in the office of Senator Tom Daschle, the Senate majority leader, and she was catching up with mail that had come in on the previous Friday. The woman slit open a hand-lettered envelope that had the return address of the fourth-grade class at the Greendale School in Franklin Park, New Jer-sey. It had been sealed tightly with clear adhesive tape. She removed a sheet of paper, and powder fell out, the color of bleached bone, and landed on the carpet. A puff of dust came off the paper. It formed tendrils, like the smoke rising from a snuffed-out candle, and then the tendrils vanished.

By this time, letters containing grayish, crumbly, granular anthrax had arrived in New York City at the offices of NBC, addressed to Tom Brokaw, and at CBS, ABC, and the *New York Post*. Several people had contracted cutaneous anthrax. The death of Robert Stevens from inhalation anthrax ten days earlier had been widely reported in the news media. The woman threw the letter into a wastebasket and called the Capitol Police.

Odorless, invisible, buffeted in currents of air, the particles from the letter were pulled into the building's high-volume air-circulation system. For forty minutes, fans cycled the air throughout the Hart Senate Office Building, until someone finally thought to shut them down. In the end, the building was evacuated for a period of six months, and the cleanup cost twenty-six million dollars.

THE HAZARDOUS MATERIALS RESPONSE UNIT of the Federal Bureau of Investigation—the HMRU—is stationed in two buildings at the FBI Academy in Quantico, Virginia. When there is a serious or credible threat of bioterrorism, an HMRU team will be dispatched to assess the hazard, collect potentially dangerous evidence, and transport it to a laboratory for analysis.

Soon after the Capitol Police got the call from the woman in Senator Daschle's office, a team of HMRU agents was dispatched from Quantico. The Capitol Police had sealed off the senator's office. The HMRU team put on Tyvek protective suits, with masks and respirators, retrieved the letter from the wastebasket, and did a rapid test for anthrax—they stirred a little bit of the powder into a test tube. It came up positive, though the test is not particularly reliable. This was a forensic investigation of a crime scene, so the team members did forensic triage. They wrapped the envelope and the letter in sheets of aluminum foil, put them in Ziploc bags, and put evidence labels on the bags. They cut out a piece of the carpet with a utility knife. They put all the evidence into white plastic containers. Each container was marked with the biohazard symbol and was sealed across the top with a strip of red evidence tape. In the early afternoon, two special agents from the HMRU put the containers in the trunk of an unmarked Bureau car and drove north out of Washington and along the Beltway. They turned northwest on Interstate 270, heading for Fort Detrick, outside Frederick, Maryland.

Traffic is always bad on Interstate 270, but the HMRU agents resisted the temptation to weave around cars, and they went with the flow. It was hot and thunderstormy, too warm for October. Interstate 270 proceeds through rolling piedmont. The route is known as the Maryland Biotechnology Corridor, and it is lined with dozens of biotech firms and research institutes dealing with the life sciences. The biotech companies are housed in buildings of modest size, often covered with darkened or mirrored glass, and they are mixed in among office parks.

The office parks thinned out beyond Gaithersburg, and the land opened into farms broken by stands of brown hickory and yellow ash. White farmhouses gleamed among fields of corn drying on the stalk. Catoctin Mountain appeared on the horizon, a low wave of the Appalachians, streaked with rust and gold. The car arrived at the main gate of Fort Detrick, where an Abrams tank was parked with its barrel aimed toward downtown Frederick. A little more than a month after September 11th, Fort Detrick remained in a condition of Delta Alert, which is the highest level of alert save for when an attack is in progress. There were more guards than usual, and they were conspicuously armed with M16s and were searching all vehicles, but the HMRU car went through without a search.

The agents drove past the parade ground and parked in a lot that faces the United States Army Medical Research Institute of Infectious Diseases, or USAMRIID, the principal biodefense laboratory in the United States. USAMRIID is pronounced "you-sam-rid," but many people call it simply Rid, or they refer to it as the Institute. USAMRIID's mission is to develop defenses against biological weapons, both medicines and methods, and to help protect the population against a terrorist attack with a biological weapon. USAMRIID sometimes performs work for outside "clients"— that is, other agencies of the U.S. government. Fort Detrick was the center of the Army's germ weapons research and development until 1969, when President Richard Nixon shut down all American offensive biowarfare programs. Three years later, the United States signed the Biological Weapons and Toxin Convention, or BWC, which bans the development, possession, or use of biological weapons. The BWC has been signed by more than one hundred and forty nations, some of which are observing the treaty while others are not.

The main building of USAMRIID is a dun-colored, two-story monolith that looks like a warehouse. It has virtually no windows, and tubular chimneys sprout from its roof. The building covers seven acres of ground. There are bio-containment suites near the center of the building—groups of laboratory rooms that are sealed off and kept under negative air pressure so that nothing contagious will leak out. The suites are classified at differing levels of biosecurity, from Biosafety Level 2 to Level 3 and finally to Level 4, which is the highest, and where scientists wearing biosafety space suits work with hot agents—lethal, incurable viruses. (A bioprotective space suit is a pressurized plastic suit that covers the entire body. It has a soft plastic head-bubble with a clear faceplate, and it is fed by sterile air coming through a hose and an air regulator.) The chimneys of the building are always exhausting superfiltered and superheated sterilized air, which is drawn out of the biocontainment zones. USAMRIID was now surrounded by concrete barriers, to prevent a truck bomb from cracking open a Biosafety Level 4 suite and releasing a hot agent into the air.

The HMRU agents opened the trunk of their car, took out the biohazard containers, and carried them across the parking lot into USAMRIID. In a small front lobby, the agents were met by a civilian microbiologist named John Ezzell. Ezzell is a tall, rangy, intense man, with curly gray hair and a full beard. FBI people who know him like to remark on the fact that Ezzell drives a Harley-Davidson motorcycle; they like his style. John Ezzell has been the anthrax specialist for the FBI's Hazardous Materials Response Unit since 1996, when the unit was formed. Over the years, he has analyzed hundreds of samples of putative anthrax collected by the HMRU. The samples had all proven to be hoaxes or incompetent attempts to make anthrax—slime, baby powder, dirt, you name it. When Ezzell was analyzing samples for the HMRU, he would often live in the USAMRIID building, sleeping on a folding cot near his lab. The agents had brought him many samples before—there had been many anthrax threats in the past. The FBI had become an important client of USAMRIID.

They went through some security doors, turned down a corridor that had green cinder-block walls, and stopped in front of the entry door to suite AA3, a group of laboratory rooms kept at Biosafety Level 3, where Ezzell worked. The agents formally transferred the containers to USAMRIID, and they gave Ezzell some chain-of-custody forms, or "green sheets," which had to be kept with the evidence, in case it was used in a trial.

Ezzell carried the containers into a small changing room at the entrance of the suite. He stripped down to his skin and put on green surgical scrubs but no underwear. He put on surgical gloves and sneakers and booties, he gowned up, and he fitted a respirator over his nose and mouth. Ezzell has been immunized to anthrax—all laboratory workers at Rid get booster shots once a year against anthrax. He carried the containers into a warren of labs in suite AA3 and placed them inside a laminar-flow hood—a glass safety cabinet with an open front in which the air is pulled up and away from a sample, preventing contamination.

Ezzell broke the evidence tape, opened the containers and the bags, and carefully unwrapped the aluminum foil. A silky-smooth, pale tan powder started coming off the foil and floating into the air, and up into the hood. The envelope inside one foil packet contained about two grams of the powder—enough to fill one or two sugar packets. It was postmarked Trenton, New Jersey, October 9th.

He opened the other foil packet, which contained the letter that had been inside the envelope. It was covered with words written in block capitals:

> 09-11-01
> YOU CAN NOT STOP US.
> WE HAVE THIS ANTHRAX.
> YOU DIE NOW.
> ARE YOU AFRAID?
> DEATH TO AMERICA.
> DEATH TO ISRAEL.
> ALLAH IS GREAT.

John Ezzell took up a metal spatula—a sort of metal knife—and slid it very slowly inside the envelope. He took up a small amount of the powder on the tip of the spatula, lifted it out, and held it up inside the hood. He wanted to get the powder into a test tube, but it started flying off the spatula, the particles dancing up and away into the hood, pulled by the current of air in the hood. The powder had a pale, uniform, light tan color. It had tested positive in the rapid field test for anthrax, and it had the appearance of a biological weapon. *"Oh, my God,"* Ezzell said aloud, staring at the particles flying off his knife.

National Security

OCTOBER 16, 2001

IN THE EARLY HOURS of the day after the anthrax-laden letter was opened in Tom Daschle's office, Peter Jahrling, the senior scientist at USAMRIID, was awakened by the sound of his pager. Jahrling (his name is pronounced "Jar-ling") lives in a small, split-level house in an outer suburb of Washington. The house is yellow and has a picket fence around it. Jahrling's wife, Daria, was asleep beside him, and their children were asleep in their rooms—two daughters, Kira and Bria, and a son named Jordan, whom Peter calls the Karate Kid because Jordan is a black-belt champion. Their oldest child, a daughter named Yara, had left for college earlier that fall.

Jahrling looked at his watch: four o'clock. He put on

his glasses, and, wearing only Jockey shorts, he walked down a short hallway into the kitchen, where his pager was sitting on the counter. It indicated that the call had come from the commander's office at USAMRIID—from Colonel Edward M. Eitzen, Jr.

Jahrling called him back. "Hey, Ed, this is Peter. What's up?"

Eitzen had been awake all night. "I want you to come into the office right now." Some issues, he said, had arisen relative to the Institute's characterization of the "sample." He was being vague. "There's highly placed interest in the sample."

Jahrling realized that the sample in question was the anthrax letter that had been delivered to USAMRIID by the FBI the previous afternoon. He figured Eitzen meant that the White House had become involved, but wasn't going to say so on an open phone line. It sounded like the National Security Council of the White House had activated emergency operations.

Jahrling returned to the bedroom and dressed quickly. He put on a light gray suit that looked like it came from Sears, Roebuck, a blue and white candy-striped shirt, and a jazzy black-and-white necktie. He fitted a silver tie bar over his tie, put on brown shoes, and hung the chain holding his federal ID card around his neck.

Peter Jahrling has a craggy face, and he wears Photogray glasses with metal rims. His hair was once yellow-blond, but it is now mostly gray. When he was younger, some of his colleagues at the Institute called him "The Golden Boy of USAMRIID" because of his blond hair and his apparent luck in making interesting discoveries about lethal viruses. He has an angular way of moving his arms and legs, a gawky posture, and it gives him the look of a science geek. It is a look he has had since he was a boy. He grew up an only child, and became fascinated with microscopes and biology at a young age. He thinks of himself as shy and socially awkward, although others think of him as blunt and outspoken, and sometimes abrasive.

Jahrling got into his car—a red Mustang with the license plate LASSA 3. His scientific interest is viruses that make people bleed—hemorrhagic fever viruses—and among them is one called Lassa, a West African virus that Jahrling studied early in his career. (He uses LASSA 1, a bashed, corroded Pontiac with a vinyl roof that's shredding away in strips, for long-distance drives, because he likes its soft seats and its boatlike ride. Daria drives LASSA 2, a Jeep.) He backed out of the driveway and drove fast along exurban roads through a beautiful night. The moon was down, and the air felt like summer, though the belt of Orion, a constellation of winter, blazed in the south. He was at the Institute by five o'clock. The place was usually dead at this hour, but the letter to Congress with some powder in it had kept people in the building overnight. He went to Colonel Eitzen's office and sat down at a conference table. Ed Eitzen is a medical doctor with thinning brown hair and a square face, eyeglasses, and a straightforward, low-key way about him. He was dressed in a pale green shirt with silver eagles on the shoulder bars, and he was looking tense. He is a well-known expert in medical biodefense. He had delivered speeches at conferences on how to plan for bioterrorism; this was the real thing.

At FBI headquarters in the J. Edgar Hoover Building on Pennsylvania Avenue in Washington, the FBI's emergency operations center, known as the SIOC (the Strategic Information Operations Center), was up and running. The SIOC is a wedge-shaped complex of rooms on the fifth floor of the headquarters, surrounded by layers of copper to keep it secure against radio eavesdropping. Desks are arrayed around a huge wall of video displays, which are updated in real time. The FBI had initiated around-the-clock SIOC operations on September 11th, and now a number of desks at the center had been devoted to the anthrax attacks. Agents from the FBI's Weapons of Mass Destruction Operations Unit were stationed at the SIOC. They had set up a live videoconference link with a crisis operations center at the National Security Council. The NSC operations center is in the Old Executive Office Building, across the street from the White House. An NSC official named Lisa Gordon-Hagerty was there and running things. The federal government had gone live.

Colonel Eitzen had been hooked into the SIOC and the NSC op center all night, while John Ezzell phoned him from his lab with the results of tests he was doing on the anthrax. Since his "Oh, my God," Ezzell had been working furiously, trying to get a sense of what kind of a weapon it was. He wasn't going to be sleeping on his cot during this terror event; he wouldn't sleep anytime soon. Meanwhile, the White House people were spinning over the word *weapon*. They wanted to know what, exactly, the USAMRIID scientists meant by the terms *weapon* and *weapons-grade*, and they wanted answers fast. What is "weapons-grade" anthrax? Had the Senate been hit with a weapon?

Jahrling and Eitzen discussed what USAMRIID should say. The White House was USAMRIID's most important client. Eitzen felt that the Institute should steer away from using the words *weapon* or *weaponized* until more was known about the powder. Jahrling agreed with him, and together they came up with the words *professional* and *energetic* to describe it, and they decided to take back the word *weapon*, which was making people too nervous.

Eitzen called the national-security people to discuss the adjustment of thinking. He used an encrypted telephone—

a secure telephonic unit, or STU (pronounced "stew") phone. A stew phone makes you sound like Donald Duck eating sushi. Eitzen said, "I'm going secure." Then, speaking slowly, he told the national-security people and the FBI what John Ezzell was learning about the anthrax.

AT SIX O'CLOCK that morning, Peter Jahrling went into his office to check his e-mail. Jahrling's office is small and windowless, and is decorated with heaps of paper along with memorabilia from his travels—a license plate from Guatemala, where he once worked as a virus hunter; a carved wooden cat; a map of Africa showing the types of vegetation on the continent; a metal telephone, with a speaking horn, that he picked up at Vector, the Russian State Research Center of Virology and Biotechnology, in Siberia. In the nineteen eighties and early nineties, the Soviets had carried out all kinds of secret work on virus weapons at Vector. The metal telephone once sat inside a clandestine Level 4 biocontainment lab; you could shout into the speaking horn while you were wearing a protective space suit—to call for help during an emergency with a military strain of smallpox, perhaps. Jahrling had been to Vector many times. He worked in the Cooperative Threat Reduction Program, which gave money to former Soviet bioweaponeers in the hope of encouraging them to do peaceful research, so they wouldn't sell their expertise to countries such as Iran and Iraq.

Jahrling sat down at his desk and sighed. There was a landfill of papers on his desk, mostly about smallpox, and it was discouraging. On top of the heap sat a large red book with silver lettering on its cover: *Smallpox and Its Eradication*. The experts in poxviruses call it the Big Red Book, and it was supposed to be the last word on smallpox, or variola, which is the scientific name of the smallpox virus. The authors of the Big Red Book had led the World Health Organization campaign to eradicate smallpox from the face of the earth, and on December 9, 1979, their efforts were officially certified a success. The disease no longer existed in nature. Doctors generally consider smallpox to be the worst human disease. It is thought to have killed more people than any other infectious pathogen, including the Black Death of the Middle Ages. Epidemiologists think that smallpox killed roughly one billion people during its last hundred years of activity on earth.

Jahrling kept the Big Red Book sitting on top of his smallpox papers, where he could reach for it in a hurry. He reached for it practically every day. For the last two years, Jahrling had run a program that was attempting to open the way for new drugs and vaccines that could cure or prevent smallpox. Scientifically, he was more deeply involved with smallpox than anyone else in the world, and he regarded smallpox as the greatest biological threat to human safety. Officially, the smallpox virus exists in only two repositories: in freezers in a building called Corpus 6 at Vector in Siberia, and in a freezer in a building called the Maximum Containment Laboratory at the Centers for Disease Control in Atlanta. But, as Peter Jahrling often says, "If you believe smallpox is sitting in only two freezers, I have a bridge for you to buy. The genie is out of the lamp."

Peter Jahrling has a high-level national-security clearance known as codeword clearance, or SCI clearance, which stands for Sensitive Compartmentalized Information. Access to SCI, which is sometimes termed ORCON information ("originator controlled"), is available through code words. If you have been cleared for the ORCON code word, you can see the information. The information is written on a document that has red-slashed borders. You look at the information inside a secure room, and you cannot walk out of the room with anything except the memory of what you've seen.

Around the corner from Jahrling's office is a room known as the Secure Room, which is always kept locked. Inside it there is a stew phone, a secure fax machine, and several safes with combination locks. Inside the safes are sheets of paper in folders. The sheets contain formulas for biological weapons. Some of the weapons may be Soviet, some possibly may be Iraqi, and a number of the formulas are American and were developed at Fort Detrick in the nineteen sixties, before offensive bioweapons research in the United States was banned. When the old biowarfare program was at its peak, an Army scientist named William C. Patrick III led a team that developed a powerful version of weaponized anthrax. Patrick held several classified patents on bioweapons.

There is probably a piece of paper sitting in a classified safe at USAMRIID—I have no way of knowing this for certain—containing a list of the nations and groups that the CIA believes either have clandestine stocks of smallpox or are trying actively to get the virus. At the top of the list would be the Russian Federation, which seems to have secret military labs working on smallpox weapons today. The list would also likely include India, Pakistan, China, Israel (which has never signed the Biological Weapons and Toxin Convention), Iraq, North Korea, Iran, the former Yugoslavia, perhaps Cuba, perhaps Taiwan, and possibly France. Some of those countries may be doing genetic engineering on smallpox. Al-Qaeda would be on the list, as well as Aum Shinrikyo, a Japanese religious cult that released sarin nerve gas in the Tokyo subway system. There is most likely a fair amount of smallpox loose in the world. The fact is that nobody knows where all of it is or what, exactly, people intend to do with it.

Having been professionally obsessed with smallpox for years, Peter Jahrling couldn't help thinking about what would happen if a loose pinch of dried variola virus had found its way into the letter to Senator Daschle. We don't really know *what* is in that powder, he said to himself. What if it's a Trojan horse? Anthrax does not spread as a contagious disease—you can't catch anthrax from someone who has it, even if the victim coughs in your face—but smallpox could spread through North America like wildfire. Jahrling wanted someone to look at the powder, and fast. He picked up his telephone and called the office of a microscopist named Tom Geisbert, who worked on the second floor. He got no answer.

TOM GEISBERT drove in that morning from Shepherdstown, West Virginia, where he lives, and arrived at the USAMRIID parking lot around seven o'clock. He was driving a beat-up station wagon with dented doors and body rust and an engine that had begun to sound like an outboard motor. He had a new pickup truck with a V-8, but he drove the clunker to save money on gas. Geisbert, who was then thirty-nine years old, grew up around Fort Detrick. His father, William Geisbert, had been the top building engineer at USAMRIID and had specialized in biohazard containment. Tom became an electron microscopist and a space-suit researcher. Geisbert is an informal, easygoing person, with shaggy, light brown hair, blue eyes, rather large ears, and an athletic frame. He likes to hunt and fish. He usually wears blue jeans and snakeskin cowboy boots; in cold weather, he'll have on a cable-knit sweater.

Geisbert went up a dingy stairwell to his office on the second floor of Rid. The office is small but comfortable, and it has one of the few windows in the building, which gives him a view across a rooftop to the slopes of Catoctin Mountain. He sat at his desk, starting to get his mind ready for the day. He was thinking about a cup of coffee and maybe a chocolate-covered doughnut when Peter Jahrling barged in, looking upset, and closed the door. "Where the heck have you been, Tom?"

Geisbert hadn't heard anything about the anthrax letter. Jahrling explained and said that he wanted Geisbert to look at the powder using an electron microscope, and to do it immediately. "You want to look for anything unusual. I'm concerned that this powder could be laced with pox. You also want to look for Ebola-virus particles. If it's got smallpox in it, everybody's going to go around saying, 'Hey, it's anthrax,' and then ten days later we have a smallpox outbreak in Washington."

Geisbert forgot about his doughnut and coffee. He went downstairs to some windows that look in on suite AA3, where John Ezzell was still working with the Daschle letter. Geisbert banged on the window and got his attention. Speaking through a port in the glass, he asked if he could have a bit of the powder to look at.

THE
DREAMING
DEMON

The Man in Room 151

ON THE LAST DAY of December 1969, a man I will call
Peter Los arrived at the airport in Düsseldorf, West Germany, on a flight from Pakistan. He had been ill with hepatitis in the Civil Hospital in Karachi and had been
discharged, but he wasn't feeling well. He was broke and
had been holed up in a seedy hotel in a Karachi slum. His
brother and father met him at the airport—his father was
a supervisor in a slaughterhouse near the small city of
Meschede, in the mountains of North-Rhine Westphalia,
in northern Germany.

Peter Los was twenty years old, a former apprentice
electrician with no job who had been journeying in pur-
suit of dreams that receded before him. He was tall and
good-looking—thin now—with a square, chiseled face
and dark, restless, rather guarded eyes under dark eye-
lashes. He had short, curly hair, and he wore faded jeans.
He was traveling with a backpack, in which he'd tucked
brushes, pencils, paper, and a set of watercolor paints, and
he carried a folding easel.

Peter Los is alive today in Germany. The details of his
character have been forgotten by the experts, but his case
and its aftermath haunt them like the ruins of a fire.

Los had been living in a commune in the city of

Bochum while he studied to be an electrician, but the members of the commune had split ideologically. Some favored a disciplined approach to communal living, while others, including Peter, favored the hippie ideals of the sixties. In August 1969—the month of the Woodstock music festival—eight members of the Bochum commune, including Peter, packed themselves into a Volkswagen bus and set off for Asia on an *Orientreise*. There were six men and two women on the bus, and they were apparently hoping to find a guru in the monasteries of the Himalayas, where they could meditate and seek a higher knowledge, and possibly also find good hashish. They drove the bus down through Yugoslavia to Istanbul, crossed Turkey, and went through Iraq and Iran, camping out under the stars or staying in the cheapest places. They rattled across Afghanistan on the world's worst roads, and the Volkswagen bus made it over the Khyber Pass. They hung out in Pakistan, but things didn't go as well as they had hoped, and they didn't connect with a guru. The two women lost interest in the trip and went back to Germany, and toward December, three men in the group drove the Volkswagen into India and down the coast to Goa, to attend a hippie festival called the Christmas Paradise. Peter stayed behind in Karachi, and ended up languishing with hepatitis in the Civil Hospital.

An eastbound train took Peter and his father and brother out of Düsseldorf, and traveled through the industrial heart of northern Germany, past seas of warehouses and factories made of brown brick. It is unlikely that Peter would have had much to say to his father at this point. He would have lit a cigarette and looked out the window. The train arrived at the Ruhr River, and it followed the course of the river into the fir-clad mountains of the Sauerland, winding upstream under skies the color of carbon steel, until it reached Meschede.

Meschede is a cozy place, where people know one another. It nestles in a valley at the headwaters of the Ruhr, beside a lake. It had been snowing in Meschede, and the hills and mountains surrounding the city were cloaked in snowy firs. It was New Year's Eve. Peter and his family celebrated the new decade, and he caught up with old friends and rested, recovering from his illness.

The weather was cloudy and dark, but in the second week of January the clouds broke away from the mountains, and clear air poured down from the north, bringing dry cold and blue skies. At the same time, influenza broke out in the town, and many people became sick with coughs and fevers. Around Friday, January 9th, Peter began to feel strange.

He was tired, achy, restless, and by the end of the day he was running a temperature. Then, on Saturday, his fever spiked upward, and he was very sick in the night. On Sunday morning, his family called an ambulance, and he was taken to the largest hospital in town, the St. Walberga Krankenhaus. He brought his art supplies and his cigarettes with him.

Dr. Dieter Enste examined Peter. He was recovering from his hepatitis, but perhaps he had typhoid fever, which is contagious, and which he could have caught in the hospital in Pakistan. They placed him in the isolation ward, in a private room, Room 151, and they started him on tetracycline.

The St. Walberga Hospital was staffed by the Sisters of Mercy, who served as nurses. The hospital was spare, simple, neat, and spotlessly clean. The isolation ward took up the entire first floor of the south wing, which was a semidetached building, three stories tall, covered with brown stucco, with a staircase that ran through the middle. The nuns told Peter to keep his door closed and not to leave his room for any reason.

He settled in on that Sunday morning and quickly began to feel better, and his fever almost went away. Even so, the nuns forbade him to leave the room, not even to use the bathroom, though it was directly across the hall. They made him use a bedpan, and they emptied it for him, and he washed himself at the sink in his room. The steam radiator under the window hissed and banged, and it made his room feel stuffy. He wanted a cigarette. He slid open one of the room's casement windows just a crack, got out his cigarettes, and lit one. The nuns were not happy with that, and ordered him to keep his window closed.

That Sunday, a Benedictine priest named Father Kunibert made rounds through the hospital, offering holy communion to the sick. He was an older man, not strong on his legs, and he worked his way down through the building, so that he wouldn't have to climb stairs. On the first floor at the end of the corridor, he put his head in Room 151 and asked the patient if he wished to receive communion. The young man was not interested. The medical report informs us that he "refused communion" and that "the priest was advised that his services were not desired."

When the nuns weren't looking, Peter continued to smoke, with his window open a crack. Cold air would pour in, filling the room with a brisk scent of the outdoors mixed with chirps of sparrows.

The tetracycline wasn't working, so the doctors started him on chloramphenicol. He had a sense of creeping malaise, an anxious feeling that things weren't right, that the drugs weren't working on his typhoid. He was restless, couldn't get comfortable, and he took out his colors and his brushes and began to paint. When he became tired of that, he sketched with a pencil. There wasn't much to see out his window—a nursing sister in a white habit hurrying down a walkway, patches of snow, branches of bare beech trees crisscrossing a sky of cobalt blue.

MONDAY AND TUESDAY PASSED. Every now and then a nun would come in and collect his bedpan. His throat was red, and he had a cough, which was getting worse. The back of his throat developed a raw feeling, and he sketched and painted. At night, he may have suffered from dreadful, hallucinatory dreams.

The inflamed area in his throat was no bigger than a postage stamp, but in a biological sense it was hotter than the surface of the sun. Particles of smallpox virus were streaming out of oozy spots in the back of his mouth and were mixing with his saliva. When he spoke or coughed, microscopic infective droplets were being released, forming an invisible cloud in the air around him. Viruses are the smallest forms of life. They are parasites that multiply inside the cells of their hosts, and they cannot multiply anywhere else. A virus is not strictly alive, but it is certainly not dead. It is described as a life-form. There was a cloud of amplified virus hanging in Room 151, and it was moving through the hospital. On Wednesday, January 14th, Peter's face and forearms began to turn red.

Stripper

THE RED AREAS spread into blotches across Peter Los's face and arms, and within hours the blotches broke out into seas of tiny pimples. They were sharp feeling, not itchy, and by nightfall they covered his face, arms, hands, and feet. Pimples were rising out of the soles of his feet and on the palms of his hands, and they were coming up in his scalp and in his mouth, too. During the night, the pimples developed tiny, blistery heads, and the heads continued to grow larger. They were rising all over his body, at the same speed, like a field of barley sprouting after rain. They were beginning to hurt dreadfully, and they were enlarging into boils. They had a waxy, hard look, and they seemed unripe. His fever soared abruptly and began to rage. The rubbing of pajamas on his skin felt like a roasting fire. He was acutely conscious and very, very scared. The doctors didn't know what was wrong with him.

By dawn on Thursday, January 15th, his body had become a mass of knob-like blisters. They were everywhere, all over, even on his private parts, but they were clustered most thickly on his face and extremities. This is known as the centrifugal rash of smallpox. It looks as if some force at the center of the body is driving the rash out toward the face, hands, and feet. The inside of his mouth and ear canals and sinuses had pustulated, and the lining of the rectum may also have pustulated, as it will do in severe cases. Yet his mind was clear. When he coughed or tried to move, it felt as if his skin were pulling off his body, that it would split or rupture. The blisters were hard and dry, and they didn't leak. They were like ball bearings embedded in the skin, with a soft, velvety feel on the surface. Each pustule had a dimple in the center. They were pressurized with an opalescent pus.

The pustules began to touch one another, and finally they merged into confluent sheets that covered his body, like a cobblestone street. The skin was torn away from its underlayers across much of his body, and the pustules on his face combined into a bubbled mass filled with fluid, until the skin of his face essentially detached from its underlayers and became a bag surrounding the tissues of his head. His tongue, gums, and hard palate were studded with pustules, yet his mouth was dry, and he could barely swallow. The virus had stripped the skin off his body, both inside and out, and the pain would have seemed almost beyond the capacity of human nature to endure.

When the Sisters of Mercy opened the door of his room, a sweet, sickly, cloying odor drifted into the hallway. It was not like anything the medical staff at the hospital had ever encountered before. It was not a smell of decay, for his skin was sealed. The pus within the skin was throwing off gases that diffused out of his body. In those days, it was called the foetor of smallpox. Doctors today call it the odor of a cytokine storm.

Cytokines are messenger molecules that drift in the bloodstream. Cells in the immune system use them to signal to one another while the immune system mounts a response to an attack by an invader. In a cytokine storm, the signaling goes haywire, and the immune system becomes unbalanced and cracks up, like a network going down. The cytokine storm becomes chaotic, and it ends with a collapse of blood pressure, a heart attack, or a breathing arrest, along with a stench coming through the skin, like something nasty inside a paper bag. No one is certain what happens in the cytokine storm of smallpox. The virus is giving off unknown proteins that jam the immune system and trigger the storm, like jamming radar, which allows the virus to multiply unhindered.

In 1875, Dr. William Osler was the attending physician in the smallpox wards of the Montreal General Hospital. He called the agent that caused the sweet smell of smallpox a "virus," which is the Latin word for poison. In Osler's day, no one knew what a virus was, but Osler knew the smell of this one. When there were few or no pustules on the skin, he would sniff at a patient's wrists and forehead, and he could smell the foetor of the virus, and it helped him nail down the diagnosis.

Around midday on Thursday, January 15th, five days after Peter Los had been admitted to the hospital, the doctors began to suspect that he had *die Pocken*—smallpox. Smallpox causes different forms of disease in the human body. Peter had classical ordinary smallpox.

The scientific name for smallpox is variola, a medieval Latin word that means "blotchy pimples." The name was given to the disease around A.D. 580 by Bishop Marius of Avenches, in the Vaud region of Switzerland. The English doctor Gilbertus Anglicus described the basic forms of smallpox disease in 1240. The virus is an exclusively human parasite. Smallpox virus can naturally infect only *Homo sapiens*. It comes in two natural subspecies, variola minor and variola major. Minor is a weak strain that was first identified by doctors in Jamaica in 1863, and is also called alastrim. While it causes people to pustulate, for some reason it rarely kills. Variola major kills around twenty to forty percent of infected humans who are not immune to it, depending on the circumstances of the outbreak and how virulent, or hot, the strain is. As a generality, doctors say that smallpox kills one out of three people.

Virus particles are also known as virions. Smallpox virions are very small. About one thousand of them would span the thickness of a human hair. It may be that you can catch smallpox if you inhale three to five infectious virions, or particles. No one knows the infectious dose of smallpox, but experts believe it is quite small.

Dieter Enste and the other doctors had not considered the idea that Peter Los might have smallpox because the young man had no rash for several days, and he had gotten a vaccination just before he had left Germany. He had gotten a second vaccination when he was in Turkey, but his vaccinations had not taken—he had not developed a scar on his arm, which meant that he had not become immune.

The St. Walberga doctors took a scalpel, cut a pustule on his skin, and drained a little of the opalescent pus onto a swab. They put it in a test tube, and a state official got in a Mercedes and drove the pus at a hundred and twenty miles an hour along the autobahn to a laboratory at the state health department in Düsseldorf.

Microscope

KARL HEINZ RICHTER was a smallpox expert in the Düsseldorf office of the state health department, a medical doctor with a kindly face and a flop of hair on one side. He wore stylish metal-framed eyeglasses and a gray sweater under a jacket, which gave him a comfy but up-to-date look. Dr. Richter, along with a team of doctors and technicians, analyzed the pus taken from Peter Los's skin. They put a little dried flake of the pus in an electron microscope—a tubelike instrument, six feet tall—which could magnify an image up to twenty-five thousand times. Then they took turns looking into the viewing hood; they would have to vote on the diagnosis.

Dr. Richter saw a vista of exploded human skin cells. Mixed in with the cellular debris were thousands of small, rounded bodies that looked like beer kegs. Some experts refer to them as bricks. The view in the microscope seemed vast, for magnified twenty-five thousand times, the flake of pus would have been an object nearly the size of a football field, and the little bricks in it lumps the size of raisins, and there could have been hundreds of thousands of them in the flake. These were virions of a poxvirus, and the vote was unanimous: this was smallpox.

The pox bricks had a crinkly, knobby surface, rather like

a hand grenade—some experts call this feature the mulberry of pox. (A mulberry is a small fruit, the size of a thumbnail, which looks like a blackberry.) There are many species and families of poxviruses; smallpox is an orthopox, a poxvirus of animals. Poxviruses are among the largest and most complicated viruses in nature. A pox particle itself either makes or consists of around two hundred different kinds of protein, and many of the proteins are locked together into the particle like a Chinese puzzle. Pox scientists are slowly picking apart the structure of the mulberry of pox, but so far nobody has figured out the full design. Experts in pox find the pox virion mathematical in its structure and almost breathtakingly beautiful. At the center of the mulberry there is an odd shape that looks like a dumbbell, which scientists call the dumbbell core or the dogbone of pox. Inside the dumbbell, or dogbone, there is a clump of DNA, which is the long, twisted, ladderlike molecule that contains the genome of smallpox—the complete blueprint and operating software for variola. The steps of the ladder of DNA are the letters of the genetic code. The genome of smallpox has about 187,000 letters, which is one of the longest genomes of any virus. Smallpox uses a lot of this code to defeat the immune system of its human host. It has about two hundred genes (which make the virus's two hundred proteins). By contrast, the AIDS virus, HIV, has only ten genes. In terms of the natural design of a virus, HIV has a simple design that works well. HIV is a bicycle, while smallpox is a Cadillac loaded with tail fins and every option in the book.

Poxviruses are one of the few kinds of viruses that are just large enough to be seen in the best optical microscopes (in which they look like fine grains of pepper). The infinitesimal palaces of biology extend far into the unseen. It is hard for the mind to grasp just how small is small in the microscopic universe of nature, but one way is to imagine a scale of nature built on the scale of the Woodstock music festival, which took place in a natural amphitheater at Max Yasgur's farm in Bethel, New York. It held up to a half-million people. Seen from low orbit above the earth, the crowd of people at Yasgur's farm would have looked something like this:

●

If a cell from the human body, in its natural size, were placed on this representation of the Woodstock festival, the cell would be an object about the size of a Volkswagen bus parked at the real festival. Bacterial cells are smaller than the cells of animals. If a single cell of E. coli (the main type of bacteria that lives in the human gut) were placed on the Woodstock on this page, it would be an object the size of a smallish watermelon, perhaps sitting on the grass beside the Volkswagen bus. A spore of anthrax would be an orange. On that same scale, a particle of smallpox would be a mulberry. (The particles of the common cold are the smallest virus particles found in nature; a cold virus would be a marijuana seed under the seat of the Volkswagen bus parked at Woodstock.) Three to five mulberries of smallpox floating into the air out of the Woodstock dot on the page would be invisible to the eye and senses, yet they could start a global pandemic of smallpox.

AS DR. RICHTER pondered the view in the microscope, he was not unprepared for the national emergency it implied. Three years earlier, he had laid out a plan for what would be done if smallpox broke out on his watch. Now it was happening. He lined up an older pox expert, Dr.

Josef Posch, and they were joined by another colleague, Professor Helmut Ippen. They organized a quarantine at the hospital, they got vaccine ready, and they gathered biohazard equipment, which Richter had previously stockpiled. He also made a telephone call to the offices of the Smallpox Eradication Program at the World Health Organization (WHO) in Geneva, Switzerland, asking for help.

The WHO occupies a building constructed in the nineteen fifties on a hill above Geneva. It is surrounded by the flags of the world's nations. In 1970, the Smallpox Eradication Program (SEP) was a relatively new effort at the WHO—it was inaugurated in 1966. The smallpox program operated out of a cluster of tiny cubicles on the sixth floor—the cubicles were exactly four feet wide, but they had a magnificent view southward across Lake Geneva toward Mont Blanc. Although the cubicles of the smallpox program were tiny and jammed together, the unit had a deserted feel, because at any given time more than half of the staff members were away, dealing with smallpox in various parts of the earth.

Dr. Richter ended up talking with an American doctor on the staff named Paul F. Wehrle, who spoke a little German. Dr. Wehrle (his name sounds like *whirly*) was a tall, thin, courtly epidemiologist with brown hair and green eyes who had a habit of wearing a jacket and tie with a white shirt when he went into the field, because he felt that a well-dressed doctor would inspire confidence in the midst of the shit terror of a smallpox outbreak. Wehrle now lives in quiet retirement with his wife in Pasadena. "I have unfortunately turned eighty," he remarked to me, "but fortunately I have all of my hair, most of my teeth, and at least some of my brain."

When Dr. Richter told him what was going on in Meschede, Dr. Wehrle understood the picture only too

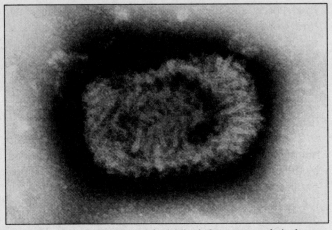

A single smallpox virus particle (virion) from a pustule in human skin. Negative contrast electron microscopy, magnified about 150,000 times, showing the "mulberry" structure of the proteins on the surface of the particle. The photograph was made in 1966 by Frederick A. Murphy, who could be described as the Ansel Adams of electron microscopy.

Diagram of a smallpox virus particle showing its surface and internal structure. Its dumbbell core (the dogbone) is visible; the dumbbell holds the genome of the virus, which consists of about 187,000 letters, or nucleotides, of DNA. (Both images courtesy of Frederick A. Murphy, School of Veterinary Medicine, University of California at Davis.)

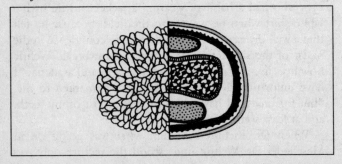

well. The WHO rule was to keep smallpox patients *out* of hospitals, because they could spread the virus all too easily—hospitals are amplifiers of variola. Smallpox could essentially sack a hospital, infecting doctors and nurses and patients, and from there the virus would continue out into the community and beyond. The WHO recommended keeping smallpox patients at home under the care of vaccinated relatives. Since there was nothing a doctor could do for a patient with smallpox, it was just as well to keep the patient away from doctors.

Wehrle went down the hall to a double cubicle that was occupied by a tall, assertive medical doctor named Donald Ainslie Henderson. Everyone called Henderson "D.A.," including his wife and children. D. A. Henderson was the head of the Smallpox Eradication Program. He was six feet two inches tall, with a seamed, rugged, blocky face, thick, straight, brown hair brushed on a side part, wide shoulders, big-knuckled hands, and a gravelly voice. Wehrle and Henderson discussed strategy, and Henderson made some telephone calls. The young man in the hospital at Meschede could start an outbreak across Europe. Henderson told Wehrle to go to Germany. Wehrle got a taxi to the airport, and that afternoon he was on a flight to Düsseldorf. Meanwhile, Henderson made arrangements to have one hundred thousand doses of smallpox vaccine shipped from Geneva to Germany immediately.

WHILE PAUL WEHRLE was en route to Meschede, Dr. Richter and the German health authorities got Peter Los out of the St. Walberga Hospital—fast. The police closed off the hospital, and a squad of attendants dressed in plastic biohazard suits and with masks over their faces ran inside the building and wrapped Los in a plastic biocontainment bag that had breathing holes in it. He lay in agony inside the bag. The evac team rushed him out of the building on a gurney and loaded the bag into a biosafety ambulance, and with siren wailing and lights flashing, it took him thirty miles along winding roads to the Mary's Heart Hospital in the small town of Wimbern. This hospital had a newly built isolation unit that was designed to handle extremely contagious patients. The Wimbern biocontainment unit was a one-story building with a flat roof, sitting in the middle of the woods. They placed Los on a silky-smooth plastic mat designed for burn victims, and he hovered on the edge of death. Construction crews began putting up a chain-link fence around the building.

That same day, Dr. Richter and Dr. Posch organized vaccinations for everyone at St. Walberga, patients and staff alike. They were given a special German vaccine that was scraped into their upper arms with a metal device called a rotary lancet, and then the doctors and their colleagues conducted interviews, trying to find out who had come into contact with Peter Los. Anyone who had seen Los's face was assumed to have breathed smallpox particles. Twenty-two people were taken to the Wimbern hospital and put into quarantine. Everyone who had been in the south wing of St. Walberga but had not seen Los's face was placed under quarantine inside the hospital, and they were ordered to remain there for eighteen days. Folding cots were brought in and set up in the bathrooms, where the medical staff slept. There wasn't enough room to hold everyone, so the authorities took over a nearby youth hostel and several small hotels in the mountains and put people there, too. After a hospital worker escaped from quarantine and went home to his family, the authorities boarded up the doors of St. Walberga and nailed them shut, and stationed a police cordon around the hospital.

Paul Wehrle arrived in Meschede on the evening of January 16th, having traveled by train from Düsseldorf. He was met at the station by Richter and Posch. (Richter did the driving, since Posch had lost an arm in the Second World War.) They took Wehrle to a hotel, and they stayed up most of the night, planning a quarantine and vaccination campaign. The Germans wanted to vaccinate people with the special German vaccine, but Wehrle did not trust it. It was a killed vaccine that the German government had been using for many years, but the WHO doctors believed it didn't give people much immunity. "The German vaccine had one small problem. It didn't work," Wehrle claims. "It was as close to worthless as a vaccine can be, only I couldn't say that to the Germans and live, because they tended to be a bit protective of their vaccine." He liked and respected the German experts and didn't want to offend them, but he gently urged them to give everyone at the hospital a second vaccination with the WHO vaccine. It couldn't hurt to have two vaccinations and might help, he said, and they agreed. He also persuaded them to use the WHO vaccine for the larger vaccination in Meschede.

The WHO maintained a stockpile of millions of doses of smallpox vaccine in freezers in a building in downtown Geneva they called the Gare Frigorifique—the Refrigeration Station. Much of the vaccine in the freezers had been donated to the Smallpox Eradication Program by the Soviet Union. The traditional vaccine for smallpox is a live virus called vaccinia, which is a poxvirus that is closely related to smallpox. Live vaccinia infects people, but it does not make most people very sick, though some have bad reactions to it, and a tiny fraction of them can become extremely sick and can die.

A staff member from the Gare Frigorifique drove a couple of cardboard boxes full of glass ampules of the Russian vaccine to the Geneva airport—one hundred thousand doses took up almost no space. The vaccine did not need to be kept frozen, because after it was thawed it would remain potent for weeks. Thousands of smallpox-vaccination needles were also shipped to Germany. They were a special type of forked needle called a bifurcated needle, which has twin prongs.

As quickly as possible, the German health authorities organized a mass vaccination for smallpox all around the Meschede area. This was known as a ring-vaccination containment. The smallpox doctors intended to encircle Peter Los and his contacts with a firewall of immunized people, so that the tiny blaze of variola at the center would not find any more human tinder and would not roar to life in its host species.

Meschede came to a halt. People left their jobs and homes, and lined up at schools to be vaccinated, bringing their children with them. A fear of pox—a *Pocken-angst*—spread across Germany faster than the virus. People who drove in cars with license plates from Meschede found that gas stations wouldn't serve them, nor would restaurants. Meschede had become a city of pox.

Nurses and doctors gave out the vaccine. A person who was working as a vaccinator would stand by the line of people, holding a glass ampule of the vaccine and a small plastic holder full of bifurcated needles. The vaccinator would break the neck of the ampule and shake a needle out of the holder. She would dip the needle into the vaccine and then jab it into a person's upper arm about fifteen times, making bloody pricks. You could have blood running down your arm if the vaccination was done correctly, for the bifurcated needle had to break the skin thoroughly. Each glass ampule was good for at least twenty vaccinations. As people passed in the line, a vacci-nator could do hundreds of vaccinations in an hour. Each needle was put into a container after it had been used on one person. At the end of the day, all the needles were boiled and sterilized to be used again the next day.

Each successfully vaccinated person became infected with vaccinia. They developed a single pustule on the upper arm at the site of the vaccination. The pustule was an ugly blister that leaked pus, and oozed and crusted, and many people felt woozy and a little feverish for a couple of days afterward, for vaccinia was replicating in their skin, and it is not a very nice virus. Meanwhile, their immune systems went into states of screaming alarm. Vaccinia and smallpox are so much alike that our immune systems have trouble telling them apart. Within days, a vaccinated person's resistance to smallpox begins to rise. Today, many adults over age thirty have a scar on their upper arm, which is the pockmark left by the pustule of a smallpox vaccination that they received in childhood, and some adults can remember how much the pustule hurt. Unfortunately, the immune system's "memory" of the vaccinia infection fades, and the vaccination begins to wear off after about five years. Today, almost everyone who was vaccinated against smallpox in childhood has lost much or all of their immunity to it.

The traditional smallpox vaccine is thought to offer protective power up to four days after a person has inhaled the virus. It is like the rabies vaccine: if you are bitten by a mad dog, you can get the rabies vaccine, and you'll probably be okay. Similarly, if someone near you gets smallpox and you can get the vaccine right away, you'll have a better chance of escaping infection, or if you do catch smallpox, you'll have a better chance of survival. But the vaccine is useless if given more than four to five days after exposure to the virus, because by then the virus will have amplified itself in the body past the point at which the immune system can kick in fast enough to stop it. The doctors had started vaccinating people at St. Walberga Hospital five and six days after Peter Los had been admitted. They were closing the barn door just after the horse had gone.

The incubation period of smallpox virus is eleven to fourteen days, and it hardly varies much from person to person. Variola operates on a strict timetable as it amplifies itself inside a human being.

The Student Nurse

JANUARY 22, 1970

ELEVEN DAYS AFTER Peter Los arrived at St. Walberga Hospital, a young woman who had been sleeping on a cot in one of the bathrooms woke up with a backache. She was a nursing student, seventeen years old, and I will call her Barbara Birke. She was small, slender, and dark haired, with pale skin and delicate features. She was a quiet person whom nobody knew much about, for she had been working at the hospital for only two weeks, and had been living in the nursing school dormitory while she received her training. The previous year, Barbara had been a kitchen helper in a Catholic hospital in Duisburg, where she had converted to the Catholic faith (her family was Protes-

tant), and she had set her sights on becoming a nurse. She
had spent Christmas with her family and had told her parents that she intended to become a nun, but she wanted to
finish nursing school before she made up her mind. The
Sisters of Mercy had reserved a place for her in the cloister.

Barbara Birke had never seen Los's face. She always
worked on the third floor of the hospital, and she had
been tending to a sick elderly man in Room 352, near the
head of the stairwell that went down through the middle
of the building. She had received both the German vaccine and the WHO vaccine a few days earlier.

Birke told the doctors that she wasn't feeling well, and
they saw that she had a slight temperature. They immediately gave her an intravenous dose of blood serum taken
from a person who was immune to smallpox. Smallpox-
immune serum is blood without red blood cells—a
golden liquid—and it is full of antibodies that fight the
virus. They put Birke inside a plastic bag, and she lay in the
bag while an ambulance carried her on the winding road
to Wimbern and through the fence to the isolation unit.

Barbara Birke developed a worried, anxious look,
while a reddening flush began to spread across her face,
shoulders, and arms, and on her legs. Her fever went up,
and her backache grew worse. Her skin remained
smooth, and no pustules appeared, although the reddening deepened in color. When the doctors pressed their
fingers on her skin, it turned white under the pressure,
but when they released their fingertips the blood came
rushing back in a moment, filling under the skin. The
doctors recognized this sign, and it was very bad.

I DON'T KNOW how much the doctors told Birke of
what they understood was coming. The red flush across

her face deepened until she looked as if she had a bad sunburn, and then it began to spread downward toward her torso. It was a centrifugal rash that had begun on the extremities. She developed a few smooth, scattered, red spots the size of freckles across her face and arms. More red spots began to appear closer to her middle, following the movement of the creeping flush. She was forbidden to have any visitors, and there were no telephones at Wimbern that the patients could use. She couldn't speak with her family.

The red spots began to enlarge, and there were more and more of them. They began to join together, like raindrops falling on a dry sidewalk, gradually darkening the pavement: she was starting to flood with hemorrhages beneath the skin.

Her back hurt, but the change in her skin was painless, and she prayed and tried to remain optimistic. Her skin was growing darker and soft and a little puffy. It was slightly wrinkled, like the skin of an old person.

The red spots merged and flooded together, until much of her skin turned deep red, and her face turned purplish black. The skin became rubbery and silky smooth to the touch, with a velvety, corrugated look, which is referred to as crêpe-rubber skin. The whites of her eyes developed red spots, and her face swelled up as it darkened, and blood began to drip from her nose. It was smallpox blood, thick and dark. The nursing nuns, who were wearing masks and latex gloves, dabbed gently at her nose with paper wipes and helped her pray.

Smallpox virus interacts with the victims' immune systems in different ways, and so it triggers different forms of disease in the human body. There is a mild type of smallpox called a varioloid rash. There is classical ordinary smallpox, which comes in two basic forms: the discrete type and the confluent type. In discrete ordinary smallpox, the pustules stand out on the skin as separate blisters, and the patient has a better chance of survival. In confluent-type ordinary smallpox, which Los had, the blisters merge into sheets, and it is typically fatal. Finally, there is hemorrhagic smallpox, in which bleeding occurs in the skin. Hemorrhagic smallpox is virtually one hundred percent fatal. The most extreme type is flat hemorrhagic smallpox, in which the skin does not blister but remains smooth. It darkens until it can look charred, and it can slip off the body in sheets. Doctors in the old days used to call it black pox. Hemorrhagic smallpox seems to occur in about three to twenty-five percent of the fatal cases, depending on how hot or virulent the strain of smallpox is. For some reason black pox is more common in teenagers.

The rims of Barbara Birke's eyelids became wet with blood, while the whites of her eyes turned ruby red and swelled out in rings around the corneas. Dr. William Osler, in a study of black-pox cases at the Montreal General Hospital that he saw in 1875, noted that "the corneas appear sunk in dark red pits, giving to the patient a frightful appearance." The blood in the eyes of a small-pox patient deteriorates over time, and if the patient lives long enough the whites of the eyes will turn solid black.

With flat hemorrhagic smallpox, the immune system goes into shock and cannot produce pus, while the virus amplifies with incredible speed and appears to sweep through the major organs of the body. Barbara Birke went into a condition known as disseminated intravascular coagulation (DIC), in which the blood begins to clot inside small vessels that leak blood at the same time. As the girl went into DIC, the membranes inside her mouth disintegrated. The nurses likely tried to get her to rinse the blood out of her mouth with sips of water.

In hemorrhagic smallpox, there is usually heavy bleeding from the rectum and vagina. In his study, Osler reported that "haemorrhage from the urinary passages occurred in a large proportion of the cases, and was often profuse, the blood coagulating in the chamber pot." Yet there was rarely blood in the vomit, and somewhat to his surprise Osler noticed that some victims of hemorrhagic smallpox kept their appetites, and they continued to eat up to the last day of life. He autopsied a number of victims of flat hemorrhagic smallpox and found that, in some cases, the linings of the stomach and the upper intestine were speckled with blood blisters the size of beans, but the blisters did not rupture.

At the biocontainment unit at Wimbern, the victim's deterioration occurred behind the chain-link fence, in a room out of sight. Dr. Paul Wehrle may have visited her (he thinks not), but there was nothing he could have said to her that would have helped, and nothing any doctor could do for her. He had seen hundreds of people dying of hemorrhagic smallpox, and he no longer felt there was any medical distinction among types and subtypes of the bloody form, that it was all an attempt by doctors to impose a scheme of order on something that was just a mess. By the time I spoke with him, the cases had flowed together in his mind, and he felt there was an inexorable sameness in the patients as the bleeding and shock came on. "It was perfectly horrifying," he said.

Barbara Birke remained alert and conscious nearly up to the end, which came four days after the first signs of rash appeared on her body. For some reason, variola leaves its victims in a state of wakefulness. They see and feel everything that's happening. In the final twenty-four hours, people with hemorrhagic smallpox will develop a pattern of shallow, almost imperceptible breaths, followed by a deep intake and exhalation, then more shallow breaths. This is known as Cheyne-Stokes breathing, and it can indicate bleeding in the brain. She prayed, and the nuns stayed with her. The Benedictine priest, Father Kunibert, who had offered communion to Peter Los, ended up at Wimbern himself with a mild case of smallpox. He may have given Birke her last rites. As the end approaches, the smallpox victim can remain conscious, in a kind of frozen awareness—"a peculiar state of apprehension and mental alertness that were said to be unlike the manifestations of any other disease," in the words of the Big Red Book. As the cytokine storm devolves into chaos, the breathing may end with a sigh. The exact cause of death in fatal smallpox is unknown to science.

PEOPLE WHO are coming down with smallpox often exhibit a worried look, known as the "anxious face of smallpox." A five-year-old girl named Ralitsa Liapsis, who came from a Greek family living in Meschede, got a worried look and broke with severe pustulation in the Wimbern isolation unit. She had been in a room at St. Walberga diagonally across the hall from Peter Los, suffering from meningitis, though she had never seen Los's face. Ralitsa spent eight weeks recovering from smallpox in the Wimbern unit, sobbing every day for her parents, who were forbidden to see her. The little girl shared her room with Magdalena Geise, a nursing student who had worked on the second floor and had never seen Los but had broken with severe ordinary smallpox. On the day after Barbara Birke died, Magdalena Geise lost her memory completely and blanked out for three weeks. Finally, as her scabs fell off and her mind returned, she did her best to comfort the scared little girl who was crying in the bed on the other side of the room. She did all she could for Ralitsa Liapsis. Magdalena was in Wimbern for twelve weeks, longer than anyone else, and when she emerged she had gone bald, and her face, scalp, and body were a horrendous mass of smallpox scars. She returned to work as a student nurse in the hospital, and wore a wig, but the patients were frightened by her appearance, and the doctors finally had to take her off the ward. A year later, Magdalena Geise's hair began to grow back, but it would take her ten years to get over her feelings of embarrassment about her appearance. Her religious faith helped her. Eventually, she married, had children and grandchildren, and found deep happiness and fulfillment. Her appearance today is that of a normal middle-aged woman with no disfigurement. Ralitsa Liapsis grew up and had children, and today the two women are friends.

Barbara Birke had had a friend at the hospital, another nursing student, Sabina Kunze, a tall, angular young woman with blond hair. Birke's death left an opening in the cloister, and Kunze decided to take her friend's place, and she made the vows and devoted her life to the work that she felt her friend would have accomplished had she lived. In the stories of Ralitsa, Magdalena, and Sabina, we see that the human spirit is tougher than variola.

Most of the people who broke with smallpox were patients and staff from the second and third floors of St. Walberga, and almost none of them had seen Peter Los's face. Doctors Richter and Posch, along with Wehrle, traced the spread of the virus and concluded that seventeen of the victims caught the virus directly from Los. Two other victims caught it from people who had caught it from Los. One of the people who caught it from him was a nun in a room in the cloistered corridor on the third floor. She survived, but another nun who was put in her room afterward came down with smallpox, went confluent, and died.

A man named Fritz Funke had arrived at the hospital one day to visit his sick mother-in-law, who was in the isolation ward at the same time Los was there. Funke waited a few minutes in a lobby, then put his head up to a door that was propped open a crack. The door opened onto the isolation corridor. Funke pleaded through the crack with a doctor to let him in, but the doctor forbade it. During the minute or so that Fritz Funke had held his face up to the door, he inhaled a few particles of variola. He had been vaccinated as an adult, in 1946, but his immunity had worn off, and two weeks later Funke was rushed to Wimbern inside a plastic bag. He survived a wicked case of smallpox. Today, the bioemergency planners know Fritz Funke as the Visitor, and they wonder about his case and see it as a disturbing example of variola's ability to spread easily through the air out of a hospital to a vaccinated visitor who barely poked his head into a ward. In the end, there were nineteen cases of variola after Los's, and there were four deaths.

Peter Los entered the stage of crust, in which the pustules begin to lose their pressure. They can rupture and leak, and they begin to develop into brown scabs that cover the body. During this phase, the bed linens of the victim become drenched with pus and extremely offensive. This was the most dangerous phase of the illness, for death often happens at the beginning of the crust, just as the patient seems to be turning the corner. But Los pulled through, and eventually they set a date for his release. A German television show called *Tage* found out about it and made plans to interview him, but he had no interest in being seen by millions. Two days before he was due to be discharged, he either climbed the fence or someone let him out, and he went home to his family. Eventually, he left Meschede, moved to West Berlin, and took various odd jobs there. It is said he went to Spain and lived on a houseboat for a time.

• • •

ONE COLD, dry day in April 1970 three months after Peter Los had been admitted to the hospital, an expert in aerosols from West Berlin arrived at St. Walberga, bringing with him a machine for making smoke. Doctors Wehrle, Posch, and Richter wanted to find out exactly how the virus had traveled through the hospital. The smoke man placed his machine in the middle of Los's old room and loaded it with a can of black soot. The doctors raised the window a couple of inches, in a re-creation of what Los had done when he disobeyed the nuns. They also left the door to the lobby propped open a crack, as it had been during the outbreak, when Fritz Funke had put his face up to it and come away infected with smallpox.

The smoke man switched on his machine, there was a whining sound, and a cloud of black smoke poured out of a nozzle and headed for Los's door and billowed down the hallway of the isolation ward. Paul Wehrle ran along with it. The smoke went through the cracked-open door and poured into the lobby, and from there it boiled up the stairs to the second floor and then went to the third floor. As it came out of the stairwell it drifted along the upper hallways. It got through the closed doors of the cloistered hallway on the third floor, and it sprinkled a number of sick nuns with black dust.

"The patients got more of a treatment than they'd bargained on when they went to the hospital," Wehrle said to me. "They were individually sooted with high-grade soot."

The soot had an energizing effect on the Sisters of Mercy—like a rock thrown into a hornet's nest. They began running up and down the stairs, crying out, *"Stoppt diesen Idioten aus Berlin! Schaltet seine Maschine ab!"*—"Stop this idiot from Berlin! Turn the machine off!"

The smoke man ignored them.

Meanwhile, Richter and Posch had gone outdoors and

were standing on the lawn. Wehrle heard them shouting, and he opened a window and looked out.

The smoke was seeping outdoors under the raised casement window and flowing in a thin, fanlike sheet up the walls of the hospital. Wehrle ran around and began opening the upper windows just a crack. To his amazement, the smoke came into the upper rooms from outside, having crept up the walls. Someone had contracted smallpox in each of those upper rooms. "It was quite a demonstration of physics, and it told us how the people had become infected," Wehrle recalled.

The smoke man was not at all surprised. He hardly raised an eyebrow. This is exactly what smoke does, he explained to the smallpox doctors. When there's a fire inside a building, naturally the smoke goes all through the building, and in cold weather it climbs the outside walls. Smallpox particles are the same size as smoke particles, and they behave exactly like smoke. A biological wildfire had occurred in Los's room, and the viral smoke had gotten into the upper floors of the hospital.

Today, the people who plan for a smallpox emergency can't get the image of the Meschede hospital out of their minds. It is a lesson in the way smallpox particles have a propensity to drift long distances, and in how a victim of the virus can escape notice for days in a hospital. People who are coming down with smallpox have days of early illness, when the virus is leaking into the air from their mouths but they haven't begun to develop a rash on their skin. A doctor would never suspect that such a patient had smallpox, because it looks like flu. The virus had ballooned in Meschede, going out of one man's mouth and into the bodies of many who had never seen him, most of whom had no idea of his existence until after they had become infected. Dr. Karl Heinz Richter and his colleagues had performed a remarkable feat of biodefense.

They were well prepared, they were ready to move in an instant, they had huge respect for the virus, and they had the full force of the WHO's Smallpox Eradication Program behind them. Even so, twenty percent of the people inside the south wing of the St. Walberga Hospital contracted smallpox. Eighty percent of them were on floors above Los's floor, and with the exception of Father Kunibert, not one of them had provably seen Los's face.

When epidemiologists study the spread of infectious diseases, they work with mathematical models. A key in any of these models is the average number of new people who catch the disease from each infected person. This number is technically called R-zero but more simply is called the multiplier of the disease. The multiplier helps to show how fast the disease will spread. Most experts believe that the multiplier of smallpox in the modern world—a world of shopping malls, urban centers, busy international airports, tourism, cities and nations with highly mobile populations, and above all nearly no immunity to smallpox—would be somewhere between three and twenty. That is, each person infected with smallpox might give it to between three and twenty more people. Experts disagree about this. Some feel that smallpox is hardly contagious. Others believe it would spread shockingly fast. The fact is, nobody knows what the multiplier of smallpox would be today, and there is only one way to find out. If it has a mulitplier of something between five and twenty, it will likely spread explosively, because five or fifteen or twenty multiplied by itself every two weeks or so can get the world to millions of smallpox cases in a few months, absent effective control. It has taken the world twenty years to reach roughly fifty million cases of AIDS. Variola could reach that point in ten or twenty weeks. The outbreak grows not in a straight line but in an exponential rise, expanding at a faster and faster rate. It begins as a flicker of something in the straw in a barn full of hay, easy to put out with a glass of water if it's noticed right then. But it quickly gives way to branching chains of explosive transmission of a lethal virus in a virgin population of non-immune hosts. It is a biological chain reaction.

Peter Los gave variola to seventeen people. Thus the initial multiplier of the disease was seventeen. Then the multiplier dropped dramatically under the effect of vaccinations and quarantine, and went quickly to zero. The chain reaction stopped. The human population was like a nuclear reactor, and the vaccine was a set of emergency control rods that were in place and ready to go, and were slammed into the reactor as fast as possible by doctors who knew exactly what they were doing.

"The main lesson of Meschede," Paul Wehrle said to me, "is that you have to be sure of the vaccine you are using."

DURING THE SCABBING PHASE, the survivors of the Meschede outbreak shed many small dark discs of dried brown skin. The scabs peppered their bedsheets and clothing, and were found scattered on the ground where they had walked. The scabs were the lifeboats of variola. The virus particles were nested in a protective web of clotted blood—the scabs were survival capsules raining from the bodies of now recovering and immune people. The virus could wait patiently for some time in a dry scab, in the hope of finding another nonimmune host, if *hope* is a word that can be applied to a virus. Variola encountered walls of resistant humanity extending all around it, and the ring of containment held at the headwaters and mountains of the Ruhr—variola disappeared from that place on the earth, and has not been seen there since.

TO
BHOLA ISLAND

Jumper

THOUSANDS OF YEARS AGO

SOMEWHERE BETWEEN ten thousand and three thousand years ago, smallpox jumped from an unknown animal into a person and began to spread. It was an emerging virus that made a trans-species jump into people from a host in nature. Viruses have many means of survival, and one of the most important is a virus's ability to change natural hosts. Species become extinct; viruses move on.

There is something impressive about the trans-species jump of a virus. The event seems random yet full of purpose, like an unfurling of wings or a flash of stripes as a predator makes a rush. A virus exists in countless strains, or quasi-species, that are changing all the time yet are stable as a whole; together, they make a species. The quasi-species of a virus are like the surface of a flowing rapids, buffeted and shaped by the forces of natural selection. The form of the virus is stable, even while the edges and surface of the river are ever in motion and shifting a little, and the river of the virus always seeks new outlets. If a particular strain of a virus that lives in an animal manages to invade a person, it may be able to replicate there, and it may get to someone else. If it keeps moving, the result is an unbroken chain of human-to-human transmission. The virus has opened a new channel to immortality. This is what HIV did about fifty years ago in central or west Africa, when two different types of HIV seem to have jumped out of sooty mangabey monkeys and chimpanzees, and began spreading in people. Very often, when a virus jumps species, it is particularly lethal in its new host.

There are many poxviruses in nature, and they infect species that gather in swarms and herds, circulating among them like pickpockets at a fair. There are two principal kinds of poxviruses: the poxes of vertebrates and the poxes of insects. Pox hunters have so far discovered mousepox, monkeypox, skunkpox, pigpox, goatpox, camelpox, cowpox, pseudo-cowpox, buffalopox, gerbilpox, several deer-poxes, chamoispox, a couple of sealpoxes, turkeypox, canarypox, pigeonpox, starlingpox, peacockpox, sparrowpox, juncopox, mynahpox, quailpox, parrotpox, and toadpox. There's mongolian horsepox, a pox called Yaba monkey tumor, and a pox called orf. There's dolphinpox, penguinpox, two kangaroopoxes, raccoonpox, and quokkapox. (The quokka is an Australian wallaby.) Snakes catch snakepox, spectacled caimans suffer from spectacled caimanpox, and crocodiles get crocpox. "Generally speaking, when crocodiles get crocpox, you see these bumps on them. I don't think it's particularly nasty for a croc," a poxvirus expert named Richard Moyer said to me. "My guess is that fish get poxes, but nobody's looked much for fish with pox," Moyer said.

Insects are tortured by poxviruses. There are three groups of insect poxviruses: the beetlepoxes, the butterflypoxes (which include the mothpoxes), and the poxes of flies, including the mosquitopoxes. Any attempt to get to the bottom of the insect poxes would be like trying to enumerate the nine billion names of God.

Insects don't have skin—they have exoskeletons—and

so they can't pustulate. Instead, poxviruses drive insects mad. A caterpillar that has caught a pox becomes nervous. It staggers around in circles on a leaf, agitated and losing its balance, and it can't seem to find its way. (This may be a caterpillar's version of "the anxious face of smallpox.") The caterpillar's development is interrupted, and the caterpillar keeps on growing bigger, until it is twice normal size. The virus is making its host larger—a nice way for a virus to amplify itself. Eventually the insect is transformed and destroyed, ending up as a swollen bag filled with a soup of insect guts and tiny crystalline nuggets that look like Wiffle balls. This soup is technically known as a virus melt. Each opening of each Wiffle ball in the melt ends up containing a particle of insect pox. The insect pox virions are inserted into the Wiffle balls and protrude from them like the knobs on a mine.

The caterpillar dies clinging to a leaf, and splits open, and out pours a spreading virus melt. The guts decay and are gone, leaving behind the Wiffle balls, which can persist for years in the environment. One day, a caterpillar comes along and eats the viral equivalent of a land mine, and melts down, and so it goes for hundreds of millions of years in the happy life of an insect pox.

No fossils of viruses have ever been found in rocks, so the origin of viruses is shrouded in mystery. Viruses are presumably very ancient, and may be similar to the earliest forms of life that appeared on the earth more than three and a half billion years ago. The insect poxes may have arisen in early Devonian times, long before the age of dinosaurs, when the seas teemed with sharks and armored fish, and the earth was covered with mosses and small plants, and there were still no trees, and the first insects were evolving. Some experts feel that the poxes of vertebrates could be the descendants of insect poxes. Smallpox, too, looks like the knobs on the Wiffle ball, though without the ball. Perhaps there was a trans-species jump of an insect pox into a newt some three hundred and fifty million years ago. Perhaps the knobs fell off the Wiffle ball when the pox got into the newt, and we are living with the consequences today.

At least two known midgepoxes torment midges. Grasshoppers are known to suffer from at least six different grasshopperpoxes. If a plague of African locusts breaks out with locustpox, the plague is hit with a plague, and is in deep trouble. Poxviruses keep herds and swarms of living things in check, preventing them from growing too large and overwhelming their habitats. Viruses are an essential part of nature. If all the viruses on the planet were to disappear, a global catastrophe would ensue, and the natural ecosystems of the earth would collapse in a spectacular crash under burgeoning populations of insects. Viruses are nature's crowd control, and a poxvirus can thin a crowd in a hurry. For most of human history, the human species consisted of small, scattered groups of hunter-gatherers. The human species did not collect in crowds, and so it was almost beneath the notice of a pox.

With the growth of agriculture, the human population of the earth swelled and became more tightly packed. Villages grew into towns, and towns grew into cities, and people began to live in crowds in river valleys where the land was fertile. At that point, the human species became an accident with a poxvirus waiting to happen.

Epidemiologists have done some mathematics on the spread of smallpox, and they've found that the virus needs a population of around two hundred thousand people living within fourteen days of travel from one another or the virus can't keep its life cycle going, and it dies out. Those conditions did not occur until the appearance of settled agricultural areas and cities, about seven thousand years ago. Smallpox could be described as the first urban virus.

The virus's genes suggest that it was once a rodent virus. Smallpox might once have lived in a rodent that multiplied in storage bins of grain. Perhaps, perhaps not. Smallpox might be a former pox of mice, or it might be a ratpox that moved on. Maybe, maybe not. There is, however, a strong suspicion that smallpox made its trans-species jump into humans in one of the early agricultural river valleys—perhaps in the valley of the Nile, or along the Tigris and Euphrates in Mesopotamia, or in the Indus River valley, or possibly along the rivers in China. By 400 B.C., the population of China had grown to twenty-five million people, which was probably the largest and densest collection of people at that time, and they were crowded along the Yellow River and the Yangtze. Down by the river somewhere, the pox found its human lover.

The mummy of the Pharaoh Ramses V, who died suddenly as a young man in 1157 B.C., lies inside a glass case in the Cairo Museum. His body is speckled with yellow blisters on his face, forearms, and scrotum. It looks like a centrifugal rash. Pox experts would very much like to look at the soles of the pharaoh's feet and the palms of his hands, to see if there are any blisters on them, for that would be a sharp diagnostic sign of smallpox. But the pharaoh's feet are wrapped in cloth, and his hands are crossed over his chest, palms downward, and the authorities at the Cairo Museum will not allow anyone to move them. Pox experts would also like to clip out a bit of the pharaoh's skin and test it for the DNA of smallpox virus, but so far that has not been allowed either.

Another possibility for the point of contact between humans and variola is Southeast Asia around 1000 B.C. Crowded city-states were developing there. Or the original host of smallpox may have been an African squirrel that lived in a crescent of green forests that are thought to have once existed along the southern reaches of the Nile

River. The climate dried out, the forests disappeared or were cut down by people, the country turned into grasslands, and the squirrel became extinct. Variola moved on.

It is possible that variola caused the plague of Athens in 430 B.C., which killed Pericles and dealt the city a devastating blow during the opening years of the Peloponnesian War with Sparta. Variola may have caused the Antonine Plague in Rome, which seems to have been carried home by Roman legions who fought in Syria in A.D. 164. Certainly smallpox rooted itself early in people living in the river valleys of China. The Chinese worshiped a goddess of smallpox named T'ou-Shen Niang-Niang, who could cure the disease. There was another goddess, Pan-chen, to whom people prayed if a victim's skin began to darken with black pox. In A.D. 340, the great Chinese medical doctor Ko Hung gave an exact description of smallpox. He believed that the disease had first come to China "from the west," about three hundred years before his lifetime.

Variola may have caused a decline in the human population of Italy during the later years of the Roman empire, making the empire more vulnerable to collapse under barbarian attacks. (The population of Italy in late Roman times may also have been gutted by malaria, or perhaps by a double whammy of malaria plus smallpox.) Variola dwelled along the Ganges River in India for at least the past two thousand years. The Hindu religion has a goddess of smallpox, named Shitala Ma, and there are temples in her honor all over India. (*Ma* means the same in Hindi as it does in English—"mother.") It is hard to say whether Shitala Ma is a good goddess or a bad one, but you certainly do not want to make her mad. In ancient Japan, smallpox arrived once in a while from China and Korea, but the virus couldn't start a chain of transmission there because the population was too thin. Eventually, around A.D. 1000, the population in Japan reached four and a half million, and apparently two hundred thousand people began to live within about two weeks' travel from one another; smallpox came to live with them, and they came to think of smallpox as a demon. In A.D. 910, the Persian physician al-Razi (Rhazes) saw a lot of smallpox when he was the medical director of the Baghdad hospital. Ancient sub-Saharan Africa had a relatively scattered human population and remained largely free of smallpox, except for occasional outbreaks along the coasts, triggered by the comings and goings of traders and slavers. The more concentrated the human population, the more likely it was to be thinned regularly by variola.

In 1520, Captain Pánfilo de Narváez landed on the east coast of what is now Mexico, near Vera Cruz. His plan was to investigate the Aztec empire, which was centered in great and powerful inland cities. One of the members of Captain Narváez's landing party was an African slave who was sick with smallpox. Variola hatched from tiny spots in the man's mouth and amplified itself into a biological shockwave that ran from the seacoast back into the Aztec empire, ultimately killing roughly half of the human population of Mexico. The wave of death that came out of less than a square inch of membrane in the mouth of Captain Narváez's man went through Central America, and it boomed along the spine of the Andes, where it gobsmacked the Inca empire. By the time the Spanish conquerors entered Peru, smallpox had softened the place up, and had killed so many people that the armies of the Incas had trouble putting up effective resistance. Smallpox had reduced the population of the Western Hemisphere while showing itself to be the most powerful de facto biological weapon the world has yet seen. (Measles was also lethal in Native American populations, and it worked alongside variola in the Americas.) During the French and Indian War, when Chief Pontiac of the Ottawa tribes was leading a siege against the British at Fort Detroit in 1763, Sir Jeffrey Amherst, the head of the British forces, wrote a letter to one of his field officers, Colonel Henry Bouquet: "Could it not be contrived to send smallpox among these disaffected tribes of Indians?" Amherst asked. "We must on this occasion use every stratagem in our power to reduce them."

Colonel Bouquet got the idea of the stratagem quite well, and his reply was to the point: "I will try to inoculate the [buggers] with some blankets that may fall into their hands." Not long afterward, one Captain Ecuyer, a British soldier, wrote in his journal: "Out of our regard for [two visiting chiefs] we gave them two blankets and a handkerchief out of the smallpox hospital. I hope it will have the desired effect." It did, and smallpox subsequently burned through the human population of the Ohio River valley, killing considerable numbers of Native Americans. This was strategic biological warfare, and it worked well, at least from the English point of view.

Vision

IN THE LATE seventeen hundreds, the English country doctor Edward Jenner noticed that dairymaids who had contracted cowpox seemed to be protected from smallpox, and he decided to try an experiment. On May 14th, 1796, Jenner scratched the arm of a boy named James Phipps, introducing into his skin a droplet of cowpox pus that he had scraped from a blister on the hand of Sarah Nelmes, a dairy worker. He called this pus "the Vaccine Virus"—the word *vaccine* is derived from the Latin word for cow. The boy developed a single pustule on his arm, and it healed rapidly. A few months later, Jenner scratched the boy's arm with lethal infective pus that he had taken from a smallpox patient—today, this is called a challenge trial. The boy did not come down with smallpox. Edward Jenner had discovered and named vaccination—the practice of infecting a person with a mild or harmless virus in order to strengthen his or her immunity to a similar disease-causing virus. "It now becomes too manifest to admit of controversy, that the annihilation of the Small Pox, the most dreadful scourge of the human species, must be the final result of this practice," Jenner wrote in 1801.

IN 1965, Donald Ainslie Henderson was thirty-six years old and was the head of disease surveillance at the Centers for Disease Control in Atlanta, when he wrote a proposal for the eradication of smallpox in west Africa. In common with most medical authorities at the time, he didn't believe that smallpox or any other infectious disease could be eradicated from the planet, but he thought that perhaps it could be done in a region. Somehow, his proposal ended up at the White House and had an effect there. For years, the Soviets had been getting up at meetings of the World Health Assembly—the international body that approves the WHO's programs—and demanding the global eradication of smallpox, and now Lyndon Johnson decided to endorse the idea. It was a political move to help improve Soviet-American relations. Henderson was abruptly called to Washington to meet with a top official in the U.S.

Public Health Service, James Watts, who informed him that he was going to WHO headquarters in Geneva to put together such a program.

"What if I don't want to go?"

"You're ordered to go," Watts said.

"Suppose I refuse?"

"Then you will resign from government service."

Henderson assumed that the attempt to eradicate smallpox would fail in about eighteen months. He told his wife, Nana, and their three children that they were going to stay in Geneva for a little while, until the program fell apart and they could come home. They put most of their things in storage in Atlanta and arrived in Geneva on the first of November. The Henderson family settled into a bungalow near Lake Geneva, not very far from the town where variola had been given its official name in A.D. 580, and they rented a refrigerator, since D.A. felt they weren't going to be there long. They would not see their stored things for another twelve years.

The Eradication program was built on the idea that variola has one great weakness: it is able to replicate only inside the human body. People have become its only natural host. Wherever it had come from in nature, it had actually lost the ability to infect its original host, and indeed, perhaps its original host had gone extinct. Variola had no reservoir of hosts in nature in which it could hide and continue to cycle if there was an attempt to eradicate it from people.

When people were infected with vaccinia, the mild cousin of smallpox, their immune systems became able to recognize variola and fight it off. If the human species could be widely infected with vaccinia and in just the right way, then vaccinia could, in effect, supplant variola in the human host. Driven out of its host by rival vaccinia, variola would have no niche left in the ecosystems of the earth.

This was, in fact, a daring plan, since no one could claim to understand the structure of natural ecosystems, especially in microbiology, or to have a clue as to whether the strategy would really work. Nature is full of surprises. Henderson wondered, for example, if smallpox just might have a little unnoticed reservoir somewhere in rodents. If so, that would destroy the dream of eradication, for humans have never been any good at getting rid of rodents. Henderson asked a virologist named James Steele if he thought any animal anywhere could harbor smallpox. Steele answered emphatically, "No. You will not find an animal reservoir." Henderson couldn't quite believe this, and for years the eradicators searched the world for a rodent, a bird, a lizard, a newt, anything with variola. They found no animal carrier of smallpox. Variola could not even replicate in primates, the closest relatives of humans. But then, in 1968, to the surprise of the eradicators, a previously unknown virus called monkeypox was discovered in a group of captive monkeys in Copenhagen, and the virus was traced back to the African rain forest, where to this day monkeypox infects humans. Monkeypox is an emerging virus that is making trans-species jumps into people in smoldering outbreaks in the rain forests of the Congo. Monkeypox may or may not one day take the natural place of smallpox vis-à-vis the human species.

Despite the evident fact that smallpox was restricted to a single host—people—many leading biologists believed that the eradication of any virus was a hopeless task. They held the opinion that it was impossible to separate a wild microbe from the ecological web it lived in. This view was expressed in 1965 by the evolutionary biologist René Dubos, in his book *Man Adapting*. "Even if genuine eradication of a pathogen or virus on a worldwide scale were theoretically and practically possible," Dubos wrote, "the enormous effort required for reaching the goal would probably make the attempt economically and humanly unwise." His belief was sensible and rational, it was held by most biologists of the time—and it was wrong.

When the program began, the World Health Assembly set a deadline of ten years for its completion. "President Kennedy had said we could land a man on the moon in ten years," Henderson said, and so it should be possible to wipe out variola in the same amount of time. At first, the leaders of the Smallpox Eradication Program weren't sure how to go about the job. They set a goal of vaccinating eighty percent of the population of countries that harbored smallpox, but that proved to be virtually impossible. They also developed the surveillance-and-ring-vaccination containment method. They tracked outbreaks of smallpox and swooped in and vaccinated everyone in a ring around the outbreak (as they would do in Meschede in 1970), which broke the chains of transmission and snuffed out the virus in that spot.

One of the lesser-known reasons for the eradication of smallpox was the desire of the doctors to eradicate vaccinia virus along with smallpox. Vaccinia gave a fairly high rate of complications, and it could make some people very sick or kill them. About one in a million people who got the vaccine during the Eradication died of it, and a larger number of people got very sick from it. The eradicators wanted to eliminate the need for vaccination, and the way to do that was to get rid of the disease. A study done by the WHO suggested that the world was losing one and a half billion dollars a year in economic damage caused by illness and complications from the vaccine.

William H. Foege is the doctor who pioneered ring vaccination. Foege, a tall, brilliant, deeply religious man, first used ring vaccination on a wide scale in Nigeria in November 1966, as an act of desperation, because he had run out of enough vaccine to immunize everybody in the area of a major outbreak. It worked surprisingly well, and as ring vaccinations proceeded and as outbreaks were choked off by rings of immune people, the eradicators began to believe that they really could wipe smallpox from the earth. The feeling was intoxicating to the eradicators. As it became clearer that the job could be done, D. A. Henderson became uncompromising as a leader. He inspired deep loyalty and affection, and he displayed the ruthlessness of a winning general. Henderson proved to be one of the geniuses in the history of management. There were normally only about eight people at headquarters, including secretaries, yet the program was a sprawling multinational operation (hundreds of thousands of health workers eventually were on salary, either part-time or full-time), and it operated all over the world, sometimes in countries engaged in civil war. His most important task was hiring the best people and giving them clear goals. Henderson's way of firing people was to suggest to them that there were jobs that were less demanding. As he explained to me, "Unless you are in a position to be tough with people, you aren't going to go forward." Either you were marching along with D. A. Henderson or you were lying flat on your face and getting a massage with tank treads.

I once asked D. A. Henderson how he felt about his role in ending smallpox. "I'm one of many in the Eradication," he answered. "There's Frank Fenner, there's Isao Arita, Bill Foege, Nicole Grasset, Zdenek Jezek, Jock Copeland, John Wickett—I could come up with fifty names. Let alone the thousands who worked in the infected countries." Even so, Henderson was the Eisenhower of the Eradication.

John Wickett was a Canadian ski bum and computer programmer who turned up in Geneva in 1971, wanting to ski the Alps while earning a little money on the side working with computers. For some reason, D. A. Henderson hired him to eradicate smallpox. Henderson had an uncanny nose for human potential in the people he hired. Today, John Wickett is widely credited as having played a big role in the Eradication. "Eradicating smallpox was the most fun I ever had," Wickett said to me. "It was fun because we actually did it and because D.A. was behind us. He could make the bureaucracy jump. When I had a problem with some bureaucrat, I'd say, 'Do you want to talk to my boss?' And I'd hear, 'No . . . ,' and the problem would get fixed."

Strange Trip

IN THE SUMMER of 1970, a twenty-six-year-old medical doctor named Lawrence Brilliant finished his internship at Presbyterian Hospital in San Francisco. He had been diagnosed with a tumor of the parathyroid gland and was recovering from an operation, so he was not able to go on with his residency. He was living on Alcatraz Island in San Francisco Bay, where he was giving medical help to a group of Native Americans who had occupied Alcatraz in a protest. He ended up doing some interviews on television from the island, and a producer from Warner Bros. saw one of them and offered him a role in a movie. The movie was *Medicine Ball Caravan*, about hippies who go to England and end up at a Pink Floyd concert. Larry Brilliant played a doctor. ("It was such a shitty movie I don't even expect my kids to watch it," he says.) The movie also featured Wavy Gravy, one of the founders of the Hog Farm commune in Llano, New Mexico. The Hog Farm commune had recently become famous for running the food kitchen at the Woodstock festival, where they also provided security. Just before the festival, Wavy Gravy had explained to the press that security would be achieved through the use of cream pies and seltzer-filled squirt bottles.

Medicine Ball Caravan was shot first in San Francisco and then in England, and during the shooting Brilliant and Gravy became friends. ("Wavy Gravy is my best friend. I was just talking with him this morning," Brilliant said to me not long ago. "I should explain that Wavy Gravy is two things: he is an activist clown and also an endangered flavor of Ben & Jerry's ice cream.") In England, Brilliant and his wife, Girija, and Wavy and his wife, Jahanara Gravy—she's from Minnesota and is said to have been Bob Dylan's girlfriend and perhaps even the model for the "Girl of the North Country"—pondered what to do next in life. A terrible cyclone had hit the delta of the Ganges River in the Bay of Bengal, in what was then East Pakistan (now Bangladesh), and the eye of the cyclone had passed over an island named Bhola. A hundred and fifty thousand people had drowned when a tidal surge had covered the entire island. The Brilliants and the Gravys hit on the idea of buying a bus and carrying food and medicines to the devastated islanders.

"Wavy and I and our wives—who, remarkably, are still our wives—drove to Kathmandu," Brilliant said. They started with a rotten old British Leyland bus that they bought cheap in London. They painted it in psychedelic colors and filled the bus with medicine and food and a bunch of hippie friends. They bought a second bus in

Germany and equipped it similarly, and the Brilliant-Gravy bus entourage made its way slowly through Turkey and Iran. The buses wandered around Afghanistan for months, and they made it over the Khyber Pass, following the same road that Peter Los and his friends had driven a little more than a year earlier in their Volkswagen bus. The Brilliant-Gravy expedition wound slowly through Pakistan and crossed into India. Civil war had broken out between East and West Pakistan—this was the independence war of Bangladesh—and the border of Bangladesh had been closed, so they couldn't get their buses into the country. They turned northward into Nepal, and eventually the buses pulled into Kathmandu. "Wavy got sick and ended up going back to the U.S. weighing about eighty pounds," Brilliant says. The Brilliants abandoned their bus in Kathmandu and went to New Delhi, India. It seems that the Brilliants were pondering what to do next in life, and nothing was coming along.

One day, the Brilliants were at the American Express office in New Delhi collecting their mail, when they encountered a man named Baba Ram Dass. Baba Ram Dass had recently been Professor Richard Alpert of Harvard University, but he and a colleague, Professor Timothy Leary, had been kicked out of Harvard for advocating the use of LSD. Baba Ram Dass spoke glowingly of a holy man named Neem Karoli Baba, who was the head of an ashram at the foot of the Himalayas in a remote district in northern India where the borders of China, India, and Nepal come together. Girija Brilliant was captivated by Baba Ram Dass's talk of the holy man, and she wanted to meet him, though Larry was not interested. Girija insisted, and so they went. They ended up living in the ashram and becoming devotees of Neem Karoli Baba, who was a small, elderly man of indeterminate age. His only personal possession was a plaid blanket. He was a famous guru in India, and the people sometimes called him Blanket Baba. The Brilliants learned Hindi, meditated, and read the Bhagavad Gita. Meanwhile, Larry ran an informal clinic in the ashram, giving out medicines that he'd taken off the bus when they'd left it in Kathmandu. One day, he was outdoors at the ashram, singing Sanskrit songs with a group of students. Blanket Baba was sitting in front of the students, watching them sing. He fixed his eye on Brilliant and said to him in Hindi: "How much money do you have?"

"About five hundred dollars."

"What about in America? How much money do you have there?"

"I got paranoid," as Brilliant explains it, "because these Indian gurus have a reputation for ripping off their students." He answered: "I have five hundred dollars in America, too."

Blanket Baba got a sly grin and started chanting, in Hindi, "You have no money. . . . You are no doctor. . . . You have no money," and he reached forward and tugged on Brilliant's beard.

Brilliant didn't know how to answer.

Neem Karoli Baba switched to English and kept on chanting. "You are no doctor . . . U N O doctor . . . U N O doctor."

UNO can stand for United Nations Organization.

The guru was saying to his student (or so the student now thinks) that his duty and destiny—his dharma—was to become a doctor with the United Nations. "He made this funny gesture, looking up at the sky," Brilliant recalled, "and he said in Hindi, 'You are going to go into villages. You are going to eradicate smallpox. Because this is a terrible disease. But with God's grace, smallpox will be *unmulun.*' " The guru used a formal old Sanskrit word that means "to be torn up by the roots." Eradicated. The word *unmulun* comes from an Indo-European root that is at least ten thousand years old—the word is probably older than smallpox.

"So I said, 'What do I do?' And he said, 'Go to New Delhi. Go to the office of the World Health Organization. Go get your job. *Jao, jao, jao, jao.*' That means, 'Go, go, go, go.' "

Brilliant packed a few things and left the ashram that night—the guru seemed to be in a rush to "unmulate" smallpox. The trip to New Delhi took seventeen hours by rickshaw and bus. When Brilliant walked into the office of the WHO, it was nearly empty. It had just been set up, and almost no one was working there. The government of India was then headed by Indira Gandhi, and she was skeptical of the Eradication program and had not yet approved it. The first person Brilliant met was the head of the office, Dr. Nicole Grasset. A French-Swiss woman who had been raised in South Africa, she was in her forties, raven haired, and dressed impeccably. Nicole Grasset has been described as a hurricane in a Dior dress.

"I was wearing a white dress and sandals," Brilliant says. "I'm five feet nine, and my beard was something like five feet eleven, and my hair was in a ponytail down my back." Grasset had no job to offer him, so Brilliant returned to the monastery and, having not slept in at least thirty-six hours, reported back to the guru.

"Did you get your job?"

"No."

"Go back and get it."

Brilliant was half dead on his feet, but the guru was looking as if he could become angry, and Brilliant did not want to have to deal with that. So he departed for New Delhi, another seventeen-hour trip, where Grasset was a little nonplussed to see the young man again so soon and looking so haggard. But nothing had changed.

"I went back and forth between New Delhi and the ashram at least a dozen times. All my teacher kept saying was, 'Don't worry, you'll get your job. Smallpox will be *unmulun,* uprooted.'" When at the ashram, Brilliant meditated. He would assume the lotus position, shut his eyes, and utter the sacred word *Aummmmm.*

Neem Karoli Baba would notice he was meditating, and he would walk up to Brilliant, yank an apple out from under his blanket, and throw it at Brilliant's crotch. There would be a *whack!* and Brilliant's *Aumm* would turn into *Oww, God! My balls!* and he would assume the "writhing lotus" position on the floor. The guru seemed to be hinting, Brilliant says now, that he needed to stand up on his feet and get back to the WHO in New Delhi, where his job awaited.

"On one of my trips, there was this tall guy sitting in the lobby of the WHO office. He looked up and said, 'Who are you? What are you doing here?' "

"I've come to work for the smallpox program," Brilliant replied.

"There isn't much of a program here."

"My guru says it will be eradicated. Who are you?"

"I'm D. A. Henderson. I'm the head of the program."

Brilliant was surprised to see the head of the global program sitting on a chair in the lobby doing nothing in particular. He later came to feel that Henderson was a little bit like the Lion in the Narnia books by C. S. Lewis. The Lion appears at key moments in the story, and he is a powerful presence who drives everything, but often you don't see him or realize what a force he is.

Henderson, for his part, was a little put off by Brilliant's white dress and his talk of a guru predicting a wipeout of smallpox. That day, Henderson wrote a note in the employment record, "Nice guy, sincere. Appears to have gone native."

Back at the ashram, Blanket Baba kept throwing apples at Brilliant's testicles. The situation was actually rather complicated. Indira Gandhi was herself a devotee of Neem Karoli Baba, and she had visited him at the monastery, where she had bowed down to him and touched his feet and asked for his advice. Blanket Baba wanted smallpox pulled up by the roots, and he was annoyed at Mrs. Gandhi for resisting the efforts of the World Health Organization to get on with the job. In fact, Neem Karoli Baba was probably the most powerful and feared mystic among the leaders of India; many of them journeyed to touch his feet and seek advice when they assumed high office. He had advised Indira Gandhi in 1962, when China invaded Indian territory in the Himalayas not far from his ashram. He had told her not to go to war with China because, he said, the Chinese army would soon withdraw from India anyway. The Chinese did partially withdraw their army, and Blanket Baba got a reputation for being able to predict the future. Larry Brilliant's trips to New Delhi were a small part of the guru's continuing effort to help India realize its future. The uprooting of smallpox, in the view of the guru, was the duty of India and was the world's destiny.

Brilliant thought he'd increase his chances of getting a job if he looked more Western, so every time he returned to New Delhi he trimmed off some of his beard and shortened his ponytail, and he began to replace articles of clothing. He ended up with medium-long hair and a short beard, and he was dressed in a checkered polyester suit with extra-wide lapels, a thick polyester tie, and a lime green Dacron shirt. He had made himself unnoticeable, for the seventies. By that time, Nicole Grasset had decided to hire him, and D. A. Henderson agreed that he might have some potential as an eradicator. He started as a typist.

Eventually, they sent Brilliant to a nearby district to handle smallpox outbreaks, where if he got into trouble they could pull him out quickly. He saw his first cases of variola major. "You can't see smallpox and not be impressed," he said. He began to organize vaccination campaigns in villages. He would go into a village where there was smallpox, rent an elephant, and ride through the village telling people in Hindi that they should get vaccinated. People didn't want to be vaccinated. They felt that smallpox was an emanation of the goddess of smallpox, Shitala Ma, and that therefore the disease was part of the sacred order of the world; it was the dharma of people to have visitations from the disease. Brilliant haunted the temples of Shitala Ma, because inside those temples people with smallpox could be found praying and dying. He would look up the local leaders, take them to a temple, chant in Sanskrit with them, and then ask for their help in dealing with smallpox. Speaking in Hindi, he told people that his guru, Neem Karoli Baba, taught that smallpox could be wiped out: "Worship the goddess and take the vaccine," he told them.

Brilliant traveled all over India with Henderson and the other leaders of the Eradication, and they came to know one another intimately. "D.A. read nothing but war novels and books about Patton and other great generals in history," Brilliant said. "Nicole Grasset read nothing except scientific things. Bill Foege was reading philosophy and Christian literature—he's a devout Lutheran. I was reading mystical literature." They ran a fleet of five hundred jeeps. They had a hundred and fifty thousand people working for the program, mostly on very small salaries. For a year and a half, at the peak of the campaign, every house in India was called on once a month by a health worker to see if anyone there had smallpox. There were a hundred and twenty million houses in India, and Brilliant estimates that the program made almost two billion house calls during that year and a half. The Lions Club and the Rotary Club International paid huge amounts of the cost of eradicating the virus in India. "Those business guys with their lapel buttons did this amazing thing," Brilliant says.

After he helped to eradicate smallpox, Larry Brilliant did other things: he became one of Jerry Garcia's physicians; he became the founder and co-owner of the Well, a famous early Internet operation; he became the CEO of SoftNet, a software company that reached three billion dollars in value on the stock market during the wild years of the Internet; he and Girija had three children; he became a professor of epidemiology at the University of Michigan; and, along with Wavy Gravy and Baba Ram Dass, he established a medical foundation called the Seva Foundation. Today, the Seva Foundation has cured two million people of blindness in India and Nepal. Along the way, Brilliant got to know Steven Jobs through their common admiration for Neem Karoli Baba. Jobs had gone to India to become a devotee of the guru, but by then Blanket Baba had gone incommunicado (he had died), so Jobs went off to study at another ashram. "Steve Jobs was a pretty nondescript guy in India, walking around barefoot with a shaved head," Brilliant recalled. "Then he started Apple Computer. I said to him, 'Steve, why are you wasting your time with this stuff? It isn't going to go anywhere.'" Jobs later donated the first seed money to start the Seva Foundation.

"I've done a lot of things in life," Brilliant said, "but I've never encountered people as smart, as hardworking, as kind, or as noble as the people who worked on smallpox. Everything about them—D. A. Henderson, Nicole Grasset, Zdenek Jezek, Steve Jones, Bill Foege, Isao Arita, the other leaders—everything about them as people was secondary to the work of eradicating smallpox. We *hated* smallpox."

"D.A. once told me he thinks of smallpox as an entity," I said.

"An entity, yes. To me, smallpox was a she, because of the goddess. You would think of her as having secret meetings with all her generals and staff, planning attacks."

Attacks came out of nowhere. Early on, Brilliant was sent to deal with an outbreak centered in a train station in Bihar—the Tatanagar Station outbreak. He was twenty-eight years old, and Shitala Ma taught him a lesson he would never forget, for the Tatanagar outbreak blew up into the largest outbreak of smallpox in the world during the years of the Eradication, and it came as a total surprise. "I went to the train station, and I found a hundred people dying of smallpox," Brilliant said. "I started crying. Women were handing me their babies. The babies were already dead. I heard rumors of birds carrying torn-off limbs of small children. Nothing in my life prepared me for that. I went to see the district medical officer and found him standing on a ladder in his office, alphabetizing his books. The look on his face was like a deer caught in headlights. 'Don't you know what's going on?' I said to him. 'What can I do?' he said."

The virus was traveling inside people up and down the railroad line. As the people moved, so did variola. The train station was exporting cases all over India and, in fact, all over the world. Brilliant began to see what a worldwide transportation system could do to amplify the virus globally in a very short time. He centered his effort first on the train station, where he found dozens of people with smallpox climbing onto a departing train. He started yelling at the stationmaster to stop the train. He had no authority, but the train stopped. He went to the police and told them to throw up roadblocks and quarantine the city. He closed the bus station and stopped all the buses from running, and he closed the airport. "I was just an American kid yelling," he says. Nicole Grasset stepped in with her authority and political connections, and she put Brilliant in charge of the operation. It took six months of desperate work, millions of dollars, and hundreds of staffers and health workers to put down the Tatanagar outbreak of variola major. "That outbreak in the Tatanagar railway station gave rise to over a thousand more outbreaks all over the world, even in Tokyo," Brilliant said. "It is not enough to think you've cornered all but that last one case of smallpox, because that last one case can create those thousand outbreaks."

Rahima

BY 1974, smallpox was nearly gone from Asia. It had waned to a handful of cases in India and Nepal, but it was not yet finished in Bangladesh. Smallpox is a seasonal virus—it breaks out and spreads more easily when the weather is dry and cool, and it diminishes in moist, warm weather. People in Bangladesh called smallpox *boshonto*, which means "spring." In south Asia, smallpox surged upward in the early spring, which is the dry season, before the summer monsoons. The eradicators mounted especially ruthless vaccination campaigns when the virus was at its ebb. In Bangladesh, they attacked the virus as hard as possible from September to November each year, when the virus seemed almost to rest—this was like killing a vampire in its sleep.

The autumn of 1974 saw a near total victory over variola in Bangladesh. In the first week of October, only twenty-four cases of smallpox were detected in the entire country. The WHO doctors could feel the end coming, and they predicted that Eradication would occur by December.

The summer monsoon of 1974 was fierce, causing the worst floods in fifty years to hit Bangladesh, especially in areas where there were still a few cases of smallpox. The floods set people in motion, and they settled in city slums called bustees. A few of them carried smallpox with them, and by December, variola major had begun to flicker unseen through the bustees of Dhaka, the capital.

In January 1975, the government of Bangladesh decided to clear out the bustees. Bulldozers flattened several in Dhaka, and the police ordered everyone to go back to their home villages. Around one hundred thousand people streamed out of Dhaka. Every person in Bangladesh lives within a two-week travel time of every other person in Bangladesh. The biological situation there is no different from what it was in Egypt or the river valleys of China thousands of years ago. Since the incubation period of smallpox is eleven to fourteen days, some of the people who came out of the bustees were incubating variola and didn't know it, and they brought it back to their villages. In February 1975, with the coming of spring, variola major roared up in more than twelve hundred places across Bangladesh. It seemed to rise out of nowhere and everywhere, coalescing out of brushfires into a viral crown fire across the country.

The event was breathtaking in its suddenness, and it shook the eradicators to the core. The rings of containment began to fail across Bangladesh. The eradicators didn't even know where to put rings, because variola seemed to be putting rings around them. They were seeing two hundred new outbreaks of smallpox every week. A failed ring vaccination was called a containment failure. During March 1975, there were nearly a thousand containment failures. It is said today that when the rings began failing in Bangladesh in the early spring of 1975, some of the leaders of the Eradication gave up hope. They felt that they had been wrong about variola after all, that ring containment wouldn't work in the end, and that the evolutionary biologists might have been right in saying that no virus could be eradicated from nature.

The program leaders in Geneva threw everything they had into the outbreak. Eradicators streamed in from the Soviet Union, Brazil, Czechoslovakia, Egypt, Great Britain, France, Sweden, and other countries. Although he had no legal authority to do so, D. A. Henderson threatened to close down the ports in Bangladesh and cut off all shipping if the government didn't mobilize its resources and get its act together. The government of Sweden poured resources into the campaign, and OXFAM, a private charity based in Great Britain, sent large amounts of money and people. Those who arrived to help received a little bit of training and were thrown into the field. The eradicators mounted ring vaccinations across Bangladesh, and they traced cases and contacts, trying to surround the life-form, and then the summer monsoons arrived, bringing wet weather. An act of nature helped to cool the viral fire, and by the end of the monsoons of 1975 smallpox was again waning. On September 15th, in Chittagong, along the eastern side of the Bay of Bengal, a boy was found with smallpox. He was the world's last case of variola major.

They waited for two months to be sure, but there were no more reported cases. Finally, on November 14th, the program leaders in Geneva sent out a press release announcing that for the first time in human history the world was free of variola major.

• • •

THE SMALLPOX ERADICATION PROGRAM team leader in charge of Bangladesh was an American doctor named Stanley O. Foster. The day after the announcement, Stan Foster received three telexes. One came from the WHO:

> CONGRATULATIONS FOR GREATEST
> ACHIEVEMENT.

Another came from the Centers for Disease Control:

> CONGRATULATIONS ALL DELIGHTED.

There was also a third telex:

> ONE ACTIVE SMALLPOX CASE DETECTED
> VILLAGE KURALIA . . . BHOLA.

Bhola Island sits in the lower delta of the Ganges and Brahmaputra rivers, where their waters merge with the Bay of Bengal. Bhola Island was the place toward which Wavy Gravy and Larry Brilliant had set out four years before in their painted buses, hoping to help someone.

Stan Foster grabbed a shortwave radio, threw a few things into a small knapsack, and left immediately for Bhola Island, traveling alone. He went to a pier in Dhaka and boarded a decrepit paddlewheel steamer called the *Rocket,* and took a passenger cabin on the deck. The *Rocket* was three hundred feet long, and it burned coal. It was a sidewheeler that had been built in 1924, and now it was a rusted hulk, jammed with humanity, chuffing and splashing down the Ganges toward the sea. Foster leaned on the rail as the boat made its way slowly along muddy channels, passing low shores lit by distant gleams of oil lamps. A waxing moon climbed across the stars, and he turned into his berth and slept. The air developed a hint of salt, and the *Rocket* entered an estuary, and shortly after sunrise the boat arrived at the port of Berisal, the end of the line, where Foster disembarked. He boarded a smallpox speedboat—an outboard motorboat run by the Eradication program—and it took him down across a vast brown bay, dotted with wooden sailing craft. He passed canoes and lateen rigs and catboats and square riggers, with cotton sails patched with cloths of bright colors; and he came to Bhola Island. It is thirty miles long, and it then contained a million people but nothing like a city. The speedboat stopped at a pier, and Foster disembarked. He was greeted by a team of local eradicators.

The island was a sandy mudflat where rice grew in profusion. There were palm trees and banana trees, and little houses of thatch, and lots of people everywhere. Foster and the local team got into a Land Rover and headed down a rutted road. The road got too muddy for the vehicle, so they parked and walked to Kuralia. They were always in the presence of people, working in the rice fields, crowding the paths. "You can't be in private in that country," Stan Foster said to me.

Local health workers led Foster and his team to a house belonging to Mr. Waziuddin Banu, a poor man who could neither read nor write. He owned no land but worked the land for others. Banu's house had a thatched roof and walls made of woven fronds of palm.

It was dark inside Banu's house. "I go in the house," Foster said, "and I can't see any cases of smallpox. Then I see this burlap sack in the corner, with a foot sticking out. It was a little kid covered with classical pox—a moderate case, not severe." The victim was a little girl, three years old, named Rahima Banu. She was frightened of Foster, and had popped herself into the sack when he came in the door. Rahima had scabbed over, and most of her scabs had already fallen off. She had caught the virus from her uncle, a ten-year-old boy named Hares. Rahima, Hares, and a few other people with smallpox in the village had been diagnosed by an eight-year-old girl named Bilkisunnessa. She reported the cases to a local health worker, and she eventually collected a reward of sixty-two dollars from the WHO—a fortune for a girl on Bhola Island.

Stan Foster raised Dhaka on the radio and told his people that he had confirmed a case of smallpox. That night, an eradicator named Daniel Tarantola put together a large team in Dhaka, with twenty motorcycles and barrels of gasoline, and they set out for Bhola Island aboard the *Rocket*. The team organized a ring vaccination on the island, and they traced contacts, and vaccinated everybody who might have been exposed. In succeeding weeks, they searched all over the island for new cases, but they didn't find any. Now variola major was really finished on earth. The hot type of smallpox had been uprooted.

When Stan Foster was with Rahima Banu, he took a bifurcated needle and used it to gently lift six scabs from her legs and feet. He tucked them into a plastic vial that had a red top. The removal of the little girl's scabs would not have hurt much, because they were falling off anyway. Each of Rahima's scabs was a brownish crust about the size of the worn nub of a pencil eraser.

When he returned to Dhaka, Foster gave the scabs to a virologist named Farida Huq, and she confirmed they were smallpox, and then she put the vial of Rahima's scabs into a metal canister, along with a sheet of paper identifying the specimen. The canister went into a cardboard mailing tube and was sent to headquarters in Geneva. A secretary named Celia Sands handled all the smallpox samples—

largely scabs in tubes—that were sent in from the field. She opened the packages on a table in the work area in the middle of the SEP cubicles, took out the red-topped plastic tubes full of scabs, and entered the information about them into a log. She was getting smallpox boosters once a year. ("Now, when you think about how we handled the specimens, it's so different from the way it's done today," she said to me. "Nothing ever happened, though.") After she had logged and inspected them, she sent the samples on to one of two smallpox repositories, either to the CDC or to the Institute for Viral Preparations in Moscow. These two places were known as WHO Collaborating Centres. Sands alternated sending samples to one or the other, so that the Americans and the Russians would end up with roughly equal amounts of scabs.

The smallpox at the Moscow Institute was cared for by a fluffy-haired, somewhat stout pox virologist named Svetlana Marennikova. She was highly regarded among pox experts, who found her scientific ideas provocative and solid.

Rahima's six scabs ended up at the CDC, where around Christmas of 1975, a pox virologist named Joseph Esposito transferred them with tweezers into a little plastic vial, smaller than a person's pinky. With an extrafine Sanford Sharpie pen, he wrote RAHIMA on the vial, added some other identifying data, and placed the vial in the CDC's reference freezer containing smallpox strains.

The strain of variola major that came from those scabs is known today as the Rahima. All six of her scabs are said to have been used up in scientific research, but the Rahima exists, frozen in small plastic vials full of translucent white ice, which looks like frozen skim milk. The milkiness is caused by vast numbers of particles of the Rahima strain, which have been grown in virus cultures and are now suspended in the ice. The Rahima sleeps in a freezer and will never die, unless and until the human

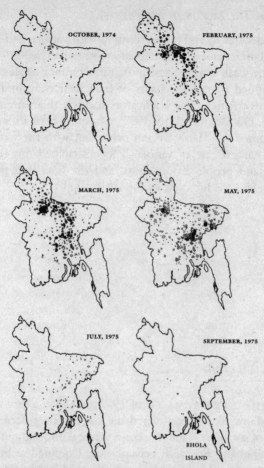

The end of variola major, autumn 1974 to autumn 1975. These successive maps of Bangladesh are like frames of a movie that show the last blowup and final eradication of variola major from the human species. You can see rings of containment around outbreaks, as well as "containment failures"—smallpox bursting out. As the virus wanes under vaccination, it moves toward the east and south, and finally it ends up on Bhola Island.

Courtesy of Stanley O. Foster, Center for Public Health Preparedness and Research, Rollins School of Public Health, Emory University, from The Eradication of Smallpox from Bangladesh, by A. K. Joarder, D. Tarantola, and J. Tulloch (New Delhi: WHO South East Asia Regional Office, 1980).

race decides to end its relationship with variola, and puts the Rahima and all other smallpox strains to death.

The weak strain of smallpox, variola minor or alastrim, continued to run in chains of transmission around the Horn of Africa. The eradicators focused their attention there. On October 27th, 1977, a hospital cook in Somalia named Ali Maow Maalin broke out with the world's final natural case of variola. They vaccinated fifty-seven thousand people around him, and the final ring tightened, and the life cycle of the virus stopped.

A Slit Throat

IN THE LATE SUMMER of 1978, less than a year after Ali Maow Maalin contracted the last naturally occurring case of smallpox, Janet Parker, a medical photographer in Birmingham, England, became sick. Confined at home, she developed a blistering rash all over her body. Her doctor believed she was having a bad reaction to a drug. Parker lived alone, and she became too ill to care for herself. Her seventy-seven-year-old father came to her house, helped Janet into his car, and drove her home to stay with him and her mother. Parker grew sicker, and her parents took her to the hospital, where doctors were stunned to discover that she had smallpox.

Mr. Parker came down with a fever twelve days after he had driven Janet home in his car, and as he was breaking with variola he died of a heart attack. Janet died of kidney failure in early September. She had been vaccinated for smallpox as an adult, twelve years before she died, but her immunity had worn off. Janet's mother broke with smallpox and survived; she was the last person on earth who is known, publicly, to have been infected with variola. In Somalia, WHO doctors described the deaths in the Parker family to Ali Maow Maalin, the hospital cook. They say he burst into tears. "I'll no longer be the last case of smallpox!" he said to them.

Janet Parker had worked in a darkroom on the third floor of a building at the medical school of the University of Birmingham. One floor below her darkroom, and down the hall some distance, a smallpox researcher named Henry Bedson was doing experiments with variola. Bedson was a thin, gentle, youthful-looking man who was internationally known and had established personal friendships with many of the eradicators. A team of investigators from the WHO was never able to pin down exactly how Janet Parker became infected, but they believed that particles of the virus had floated out of Bedson's smallpox room, drifted through a room used for animal research, had then been sucked into the building's air-vent system, had traveled upward one floor, passed through a room known as the telephone room, passed through two more small rooms, and finally gotten inside Parker's darkroom, and had lodged in her throat or lungs.

On September 2nd, as Janet Parker lay desperately ill, Henry Bedson was discovered lying unconscious in the potting shed behind his house. He had slit his throat with a pair of scissors, and much of the blood in his body had drained out. He died five days later, despite transfusions. When Bedson slit his throat, the eradicators woke up to the fact that although the disease was gone, the virus wasn't, and they stepped up their efforts to gain control of all the known stocks of smallpox in the world. They felt that as human immunity to the virus waned year by year, the potential for laboratory accidents was growing.

In 1975, at least seventy-five laboratories had frozen stocks of smallpox virus. Poxviruses, including smallpox, can survive for many decades in a freezer without damage or loss of infective potency—probably for at least fifty years. A freezer with a few vials of smallpox in it could become a biological time bomb. In 1976, a year before the last natural cases of smallpox occurred, the WHO formally asked all laboratories holding smallpox to either destroy their stocks or send them to one of the two Collaborating Centres. The WHO had no legal power to compel anyone to give up their smallpox, but D. A. Henderson and the others were tough and persistent. One by one, the laboratories that were keeping smallpox sent their samples to America and Russia or destroyed them or said they had destroyed them.

Vault

TODAY, VARIOLA EXISTS OFFICIALLY in only two repositories, the Collaborating Centres. One of the repositories is the Maximum Containment Laboratory at the CDC in Atlanta. The other repository is in Russia. When scientists handle variola, international rules require them to wear full space suits and to be inside a sealed Biosafety Level 4 containment zone. The WHO forbids any laboratory from possessing more than ten percent of the DNA of variola, and no one is officially allowed to do experiments with smallpox DNA. Variola is now exotic to the human species, highly infective in humans, lethal, and difficult or impossible to cure. It is generally believed to be the most dangerous virus to the human species.

The CDC's smallpox collection sits inside a liquid-nitrogen freezer. The freezer is a stainless-steel cylinder, about chest high, with a circular lid and a digital temperature display. At the bottom of the freezer there is a pool of liquid nitrogen, three inches deep, which maintains the air inside the freezer at a steady temperature of minus 321 degrees Fahrenheit. There are about four hundred and fifty different strains of smallpox inside the freezer. The samples are frozen in the little plastic vials called cryovials. The cryovials stand upright in small white boxes made of cardboard or plastic, which are divided with grid inserts, like cartons for storing wine. The boxes are stacked in metal racks, and they sit suspended over the pool of liquid nitrogen, bathed in cold fumes. The entire volume of the CDC's smallpox is about the size of a beach ball.

Officials at the CDC do not comment on such matters as where exactly the smallpox is stored or what the freezer looks like. The freezer is on wheels, and it can be moved around, and it may be moved from time to time, as in a shell game. It is covered with huge chains that are festooned with padlocks the size of grapefruits. The chains are connected to anchors or bolts in the floor or the walls, so that the freezer can't be moved unless the chains are unlocked or cut. I have been told that the smallpox freezer can often be found sitting inside a steel chamber that is said to resemble a bank vault. The variola vault is enlaced with alarms, and it may be disguised. You might look straight at the vault and not know that your eyes are resting on the place where half the world's known smallpox is hidden. There may be more than one variola vault. There may be a decoy vault. If you opened the decoy vault, you could find a freezer full of vials labeled SMALLPOX that held nothing but vaccine—a raised middle finger from the CDC to a feckless smallpox thief. The variola vault could be disguised to look like a janitor's closet, but if you opened the door in search of a mop, you could find yourself face-to-face with a locked vault, having set off screaming alarms. If the variola alarms go off, armed federal marshals will show up fast.

The smallpox at the CDC's repository may be kept in mirrored form: there may be two freezers, designated the A freezer and the B freezer. The A and B freezers (if they exist, which is unclear) would each contain identical sets of vials—mirrored smallpox—so that if one freezer malfunctioned and its contents were ruined, the variola mirror would remain. No one will talk about mirrored smallpox today, but twenty years ago the smallpox *was* kept in mirrored form at the CDC. Whether that arrangement holds true today is presumably not known to anyone but a handful of top people at the CDC and to some of the security staff. People at the CDC do not discuss details of the storage, and many of them may not know of the existence of the vault. They don't know, and they don't ask.

THE
OTHER SIDE
OF THE MOON

A Flash of Darkness

DR. CHRISTOPHER J. DAVIS, a British intelligence officer, was tidying up his office in the old Metropole Building off Trafalgar Square and was getting ready to catch a train home to Wiltshire at the end of a chill, dank day. Davis was an analyst on the Defence Intelligence Staff, with an area of expertise in chemical and biological weapons. He is a medical doctor with a Ph.D. from Oxford, and was then a surgeon commander in the Royal Navy. He has a serious, crisp manner, trim good looks, blue eyes and light brown hair, and an angular face.

The papers on Davis's desk contained source intelligence—bits and pieces of information, some credible, some not, about chemical and biological weapons that some countries might or might not have. His job was to take all the bits and move them around, look at them, like fragments of broken glass, and try to assemble them into a picture of something. Chemical and biological weapons were then a backwater. Christopher Davis peered into the wastebasket—you couldn't leave any papers there. Below his window, people were heading off into the darkness across Great Scotland Yard, toward their pubs and the Tube. He was anticipating with pleasure the long train ride home. . . . He could decompress, read, sleep. . . . A little trolley would come along with food. . . .

The telephone rang. It was his boss, a man referred to as ADI-53. "Chris, you'd better come to my office right away. I've got a telegram you need to look at."

Davis dropped and locked—dropped all the loose papers into combination safes in his office, spun the tumblers, locked his office—and hurried down the hall.

ADI-53 handed him a two-page, highly secret telegram. He said that the Secret Intelligence Service (SIS), which is also called MI6, was "holding a high-level chap who's just defected from the Soviet Union." The SIS guys were keeping the man in a safe house outside London. He was a fifty-three-year-old chemist named Vladimir Pasechnik, the director of the Institute for Ultrapure Biopreparations in St. Petersburg. Dr. Pasechnik had been attending a drug-industry trade fair in Paris, and he had abruptly sought asylum in the British embassy. He was a so-called walk-in, an unexpected defector. The SIS people had taken him in for an immediate debriefing, and the telegram summarized the results. It was largely in Pasechnik's own words: "I am part of Biopreparat, a large, secret program which is devoted to research, development, and production of biological weapons in the USSR," it began. Two words in the telegram jumped out at Davis. They seemed to burn on the page: *plague* and *smallpox*.

Plague is *Yersinia pestis*, a bacterial microbe widely known as the Black Death, a contagious pestilence that wiped out one third of the population of Europe around 1348. Plague can travel from person to person through the air, propelled by a pneumonialike cough.

"Oh, shit!" Davis said to his boss.

Davis realized he was looking at a strategic biowarfare program. Plague and smallpox are not tactical weapons. They can't be used in any sort of limited attack: they are designed to go out of control. They are intended to kill large numbers of people indiscriminately, and they have no other function. The target of smallpox is a civilian human population, not a concentration of military forces. At the end of the day, you can deal with anthrax because it is not transmissible in people, but plague and smallpox are entirely different matters. "If what is in front of me is accurate," Christopher Davis said to his boss, "it means that they have strategic biological weapons. It also means they have launch systems or other means of delivery. We just haven't found the systems yet."

EARLY THE NEXT WEEK, at a colorless business hotel in the south of London, Davis met Vladimir Pasechnik, who sat in a room with his handlers from MI6. They called him by his first name, and Davis became his main debriefer. Over a period of many months, he met with Vladimir in various hotels around London, listened to him, and asked questions. There were always handlers in the room, and there was always a technical specialist from the SIS. They did not bring Pasechnik into the headquarters of MI6, because it was assumed that KGB operatives had the place staked out. Vladimir had left his wife and children behind, and he was very worried about them.

He told Davis that Biopreparat, also known as the System, was huge. The program had vast stocks of frozen plague and smallpox that could be loaded onto missiles, although Pasechnik was not sure of the intended targets. The warhead material had been genetically engineered, he said. He understood only too well the modern techniques of molecular biology, as did his colleagues. One of the principal weapons was genetically modified (GM) plague that was resistant to antibiotics. The Soviet microbiologists had created this GM plague with brute-force methods: they had taken natural plague and had exposed it again and again to powerful antibiotics, and in this way, they forced the rapid evolution of drug-resistant strains. This sort of research is known among bioweaponeers as "heating up" a germ. The heated-up *pestis* would spread from person to person in a lethal cough, and doctors would not have drugs to treat it effectively. One of the strains of GM plague was being manufactured by the ton, Vladimir said. He also said that Biopreparat scientists were trying to come up with even more powerful strains using the techniques of molecular biology—inserting foreign genes into plague to further heat it up.

Vladimir said that lately the Soviet Ministry of Defense had been demanding that biologists develop a new manufacturing process for making tonnage amounts of weapons-grade smallpox. Military biologists had been using an older process for making smallpox into warhead material, and now there was a new generation of missiles that they wanted to arm with variola. The Soviet military had long considered smallpox a strategic weapon—during the Eradication, when the Ministry of Health had been making and donating vaccine to the WHO, the Ministry of Defense had been making and stockpiling smallpox as a weapon. Much of the advanced work with smallpox was now happening in Siberia, at the Vector research facility, but he didn't know much about it, he said.

Vladimir Pasechnik was anxious about the genetic-engineering research at Biopreparat. He was afraid that a genetically engineered virus or germ could escape from the weapons program. He said that genetic engineering was why he had defected. He didn't want money, he wanted out. "I couldn't sleep at night, thinking about what we were doing in our laboratories and the implications for the world," he told his British debriefers.

• • •

THE BRITISH had been sending encrypted messages to the CIA to inform them of what Pasechnik was saying, but they wanted to go face-to-face with the Americans for a comprehensive meeting. Christopher Davis and his colleagues wrapped up the debriefing of Pasechnik in the late spring of 1990. The British government then sent Davis and a close colleague from Defence Intelligence, Hamish Killip, to the headquarters of the CIA in Langley, Virginia, where they briefed their American colleagues on the details of the GM Black Death, and smallpox, and of the missiles tipped with bioweapons. The British weren't absolutely certain that the biological strategic missiles were operational and ready for launch, but if they were, it was pretty clear that they would be targeted on North America.

Several years later, Christopher Davis would receive the Order of the British Empire from Queen Elizabeth II. Though the Queen didn't know it, he had received the O.B.E. for having said "Oh, shit" to his boss—it marked the first insight into the fact that the Russian biowarfare program was strategic, like a nuclear program.

"I have the highest respect for the intelligence services of the USA," Davis said to me, recalling his visit to Langley, "yet they were amazed at what we told them." The CIA officials may also have been dismayed that British intelligence had cracked open a strategic-weapons program in Russia that they had not known very much about. In the world of intelligence, it is not good to be told something new and important by an intelligence officer from another government. Yet even while they listened to Davis and Killip, the CIA people had their own secret knowledge, which they did not share with the British. They had classified this information as NOFORN, meaning that no foreigners could have it.

Forbidden Planet

SOMETIME before 1991, a Soviet intercontinental ballistic missile was launched from Kamchatka, the peninsula that hangs down from Asia into the northern Pacific Ocean. It carried a massive MIRV (multiple independent reentry vehicle) payload. A MIRV payload separates into individual warheads, which land on discrete targets. The MIRV itself is called a bus. It is rather like a bus: it carries the warheads and lets them off to head for their separate destinations.

American spy satellites and Navy ships watched the missile as it soared out of Kamchatka and above the atmosphere. The MIRV bus detached from the launch vehicle and went on a free-fall arc through space over the Pacific Ocean. The bus separated into ten warheads, and they fell into the sea. The American sensors pulled in some data about the shot, which had to be decoded and assembled and thought about. This took time, but something strange began to emerge. There was something different about this MIRV. The bus had an unusual shape, and it did odd things as it moved through space—rather than spinning, as the usual nuclear warheads did, it oriented itself in relation to the earth. Infrared cameras on American satellites photographed something that they had never seen on a Russian warhead before: a large fin panel that was glowing with heat—the bus was dumping heat into space as the vehicle soared over the Pacific. Why would it need to do that?

The laws of thermodynamics said that if there was heat pouring into space from the bus, then the inside of the bus had to be cold. This was a refrigeration system. But what on the bus needed to be kept cold? A nuclear warhead can tolerate heat above the boiling point of water. After the bus separated into its ten small warheads, each warhead punched down through the atmosphere, popped a parachute, and fell into the water. Nuclear warheads don't need to come down on a parachute.

Several such tests took place, but it's not clear when they happened or how much information the CIA really got. Analysis takes time, and nothing is ever crystal clear. In October 1988, the CIA obtained imagery of missiles sitting in storage bunkers or launch silos in Kamchatka. The imagery showed that the warheads were connected by pipes or hoses to refrigeration systems on the ground. All the Soviet missiles used liquid fuel, which needed to be kept cold, but even so, something about these cooling systems made the CIA analysts think they were not for cooling rocket fuel. Refrigeration implies life. The missiles appeared to contain living weapons.

The CIA has close ties with British intelligence. Even so, the CIA chose not to tell MI6 about the tests of the new missile warheads. The CIA could not be absolutely certain that the warheads were biological, or that a germ or virus could possibly be powerful enough to use in place of a nuclear weapon. It seems that there was a puzzlement going on within the American intelligence community over whether or not a germ that landed on a city from space could do any kind of real damage. And yet if it was true that biological missiles seemed to be aimed at the

United States, just who should be informed of this? The NOFORN knowledge of the chilled biowarheads was tucked away inside the CIA like the meat in a walnut.

SHORTLY AFTER Christopher Davis and Hamish Killip briefed the Americans on what Dr. Pasechnik had told them, the United States and Britain became a great deal more concerned about biological weapons. President George Bush and Prime Minister Margaret Thatcher were briefed by their intelligence people on the ICBMs armed with plague and smallpox. Mrs. Thatcher hit the roof. She telephoned Mikhail Gorbachev, who was then the head of the Soviet Union, and forcefully asked him to open his country's biowarfare facilities to a team of outside inspectors. Gorbachev stalled for a while, but he eventually agreed.

A secret British-American weapons-inspection team toured four of the main Biopreparat scientific facilities in January 1991. The team members included Christopher Davis. They ran into the same problems that the United Nations inspectors would later run into in Iraq. The Soviet biologists did not want to discuss their work and did not want anyone seeing their laboratories in operation. The inspectors were met with denials, evasions, time-wasting bureaucracy, stupefying, alcohol-laden meals that stretched on for hours, snarled transportation arrangements, and endless speeches about friendship and international cooperation. Whenever they could pull themselves away from a speech, they saw large Level 4 space-suit rooms that had been completely stripped of equipment and sterilized and were not in use, though the labs showed every sign of having been in operation recently. They traveled by bus to a huge microbiology facility south of Moscow called Obolensk. The facility was surrounded by layers of barbed wire and military guards. The head scientist was a lean-faced military officer and microbiologist named Dr. Nikolai Urakov, an expert in plague. Inside one of the Level 4 areas, the inspectors found an array of two-story-tall fermenter-production tanks. This was a major production facility for the GM plague, but the tanks were now empty. When Davis and the other inspectors accused Dr. Urakov of manufacturing plague by the ton, he blandly informed the inspectors that all the research at his institute was for medical purposes, since plague was "a problem" in Russia.

"This was clearly the most successful biological-weapons program on earth, yet these people just sat there and lied to us, and lied, and lied," Davis said to me. He insists that the Russian government has never come clean. "To this day, we still do not know what happened in the military facilities that were the heart of the Russian program."

Late in the day on January 14th, the team arrived at Vector, the sprawling virology complex situated in the larch and birch forests near a town called Koltsovo, about twenty miles east of Novosibirsk, in Siberia. They were offered vodka and caviar, lots of good food and many toasts to friendship, and were sent to bed. The next morning, after being treated to more vodka and caviar for breakfast, they demanded to see the building called Corpus 6. It is a homely brick structure, with windows rimmed in concrete. The stairs in Corpus 6 are crooked. Many of the buildings at Vector were constructed by gangs of prison laborers, and it is said that they wanted to make every concrete step slightly different in size. The Russian story is that the prisoners were hoping that some biologist would fall down the stairs and break his stinking neck.

The inspectors were shown into the entry area of Corpus 6. A British inspector named David Kelly, a well-known research microbiologist at Oxford University, took a technician aside and asked him what virus they were working with there.

"We are working with smallpox," the technician answered.

By early 1991, smallpox was supposed to exist only at the CDC and at the Moscow Institute. David Kelly was amazed to hear the word *smallpox,* and he repeated the question three times—"You mean you were working with variola major here?"—and he emphasized to the technician that his answer was very important. The technician responded emphatically, three times, that it was variola major. Kelly says that his interpreter was the best Russian interpreter that the British government has. "There was no ambiguity."

The inspectors were stunned. Vector was not supposed to have any smallpox at all, much less be doing experiments with it.

The inspectors made their way up the crooked stairs of Corpus 6 to an upper level, and they entered a corridor. Along one side of the corridor was a line of glass windows looking in on a giant steel dynamic aerosol test chamber. The device is for testing bioweapons—it has no other purpose. Small explosives, or bomblets, are detonated inside the chamber, releasing a biological agent into the air of the chamber. The aerosol test chamber in Corpus 6 had tubes coming out of it. Sensors could be placed on the tubes—or monkeys or other animals could be clamped onto them—and exposed to the chamber's air. On the other side of the corridor was a command center that bespoke serious business. The center had massive dials, lights, and switches that made it look like a set from a Russian remake of *Forbidden Planet.* ("It's Krell metal. . . . Try your blaster on that, Captain.")

The Vector scientists later explained to the inspectors that the chamber was a Model UKZD-25 bioexplosion test chamber. It was the largest and most sophisticated modern bioweapons test chamber that has been found in any country. The inspectors came to believe that the bomblets for the smallpox MIRV biowarheads had probably been tested and refined in the chamber.

The inspectors asked if they could put on space suits and go inside the chamber. They would have liked to take swab samples from the inner walls, but the Russians refused. "They said our vaccines might not protect us. This suggested that they had developed viruses that were resistant to American vaccines," one of the inspectors, Dr. Frank Malinoski said. The Russians became agitated and ordered the inspectors to leave Corpus 6.

At a sumptuous dinner that evening, full of toasts to the new relationship, three inspectors—David Kelly, Frank Malinoski, and Christopher Davis—publicly confronted the head of Vector, a pox virologist and scientific administrator named Lev S. Sandakhchiev, about Vector's smallpox. (His name is pronounced "Sun-dock-chev" but many scientists refer to him simply as Lev.) He backpedaled angrily. "Lev is gnomelike, a short man with a wizened, weather-beaten, lined face and black hair," Christopher Davis said to me. "He's very bright and capable, a tough individual, full of bonhomie, but he can be very nasty when he is upset."

Sandakhchiev heatedly insisted that his technician had misspoken. He called on his deputy, Sergey Netesov, to support him. The two Vector leaders said that there had been no work with smallpox at Vector. The only place smallpox existed in Russia was at the WHO repository at the Moscow Institute. They said they had been doing genetic engineering with smallpox genes, that was all. Vector didn't have any live smallpox, they said, only the virus's DNA. The more they spoke about genetic engineering and the DNA of smallpox, the murkier and scarier the talk sounded to the inspectors. "They were both lying," David Kelly said to me, "and it was a very, very tense moment. It seemed like an eternity."

"The fact is, they had been testing smallpox in their explosion test chamber the week before we arrived," Christopher Davis said. "The nerve of these people."

The first deputy chief of research and production for Biopreparat, Dr. Kanatjan Alibekov, who was present at the Vector meetings, later defected to the United States in 1992. He became Ken Alibek, and he revealed a panoramic vista of Biopreparat, along with details that Christopher Davis and the others had not imagined. Alibek described a huge program that was broken into secret compartments. Very few people inside the program knew its scope. Because it was compartmentalized and secret, it had the potential to fall apart into smaller pieces, and the world might never know where all the pieces had gone.

IT IS NOW CLEAR that the Soviet bioweapons program was quite advanced by the time the Soviet government fell, in December 1991. A couple of years earlier, in 1989, at a military facility known as the Zagorsk Virological Center, about thirty miles northeast of Moscow, biologists were making and tending a stockpile of twenty tons of weapons-grade smallpox. This is absolutely extraordinary, considering the security arrangements that prevail around the little collection of smallpox vials in Atlanta. The Zagorsk smallpox was apparently kept in insulated mobile canisters, so that it could be moved around on railcars or in cargo aircraft. It seems that there was another stockpile of frozen smallpox warhead material at a military facility called Pokrov, about fifty miles east of Moscow.

The biowarheads, Ken Alibek revealed, could be filled with dry powder or with liquid smallpox. Each MIRV bus had ten warheads, and each warhead had ten grapefruit-sized bomblets inside it. The warheads would float toward the earth on parachutes, and as they neared the ground they would burst apart, throwing out a fan of bomblets. Each bomblet could hold two hundred grams of liquid smallpox. The bomblets were likely pressurized with carbon-dioxide gas, which blew out a mist of variola. Each warhead could deliver a half gallon of smallpox mist, hissing from the bomblets as they rained down. The mist would drift above rooftops, and it would get into people outdoors, and it would get inside houses and schools, and it would be sucked into the vents of office buildings and shopping malls. One MIRV missile could deliver forty-five pounds of smallpox mist into a city. It doesn't sound like a lot, until we consider how much smallpox Peter Los put into the air by coughing.

The smallpox that was designated for the warheads was evidently a strain that the Soviets named India-1. It had been collected in India in 1967, in a little place called Vopal, by Russian scientists who were apparently ordered by the KGB to get some really hot scabs. They probably tested this strain against other strains to get a sense of which was the hottest, or perhaps they selected a strain that seemed more resistant to vaccine. (This would almost certainly have required human testing.) In any case, the Vopal strain, or India-1, became a strategic weapon. The strain may be exceptionally virulent in humans. Officials of the Russian Federation have vaguely admitted to the existence of India-1, but the Russian government has so far refused to share the India-1 strain with any scientists outside Russia, and so its characteristics, and the means to defend against it, remain uncertain.

In 1991, the WHO had two hundred million doses of

frozen smallpox vaccine in storage in the Gare Frigorifique in downtown Geneva. This was the world's primary stockpile of smallpox vaccine. The vaccine stockpile was costing the WHO twenty-five thousand dollars a year in storage costs, largely for the electricity to run the freezers. In 1991, an advisory panel of experts known as the Ad Hoc Committee on Orthopoxvirus Infections recommended that 99.75 percent of the vaccine stockpile be destroyed, in part to save on electricity costs. Since the disease had been eradicated, there was no need for the vaccine. The vaccine was taken out of the freezers, sterilized in an oven, and thrown into Dumpsters. This move saved the WHO less than twenty-five thousand dollars a year, and left it with a total of five hundred thousand doses of smallpox vaccine. That is less than one dose of the vaccine for every twelve thousand people on earth. The WHO has no plans to increase its stockpile now, since replacing the lost quantity would cost a half-billion dollars, and it doesn't have the money.

According to several independent sources, Lev Sandakhchiev was in charge of a research group at Vector in 1990 that devised a more efficient way to mass-produce warhead-grade smallpox in industrial-scale pharmaceutical tanks. In 1994—three years after the British and American bioweapons inspectors toured Vector and were told by Sandakhchiev that there was no smallpox there— his people built a prototype smallpox bioreactor and allegedly tested it with variola major. The reactor is a three-hundred-gallon tank that looks something like a hot-water heater with a maze of pipes around it. It sits on four stubby legs inside a Level 4 hot zone in the middle of Corpus 6, on the third floor of the building. The reactor was filled with plastic beads on which live kidney cells from African green monkeys were growing. Vector scientists would pump the reactor full of cell-nutrient fluid and a little bit of smallpox. The reactor ran at the temperature of blood. In a few days, variola would spread through the kidneys cells, and the bioreactor would become extremely hot with amplified variola, whereupon the liquid inside the reactor could be drawn off in pipes and frozen. In biological terms, the liquid was hot enough to have global implications. A single run of the reactor would have produced approximately one hundred trillion lethal doses of variola major—enough smallpox to give each person on the planet around two thousand infective doses of smallpox. Vector scientists steadfastly maintain, however, that they did no experiments with smallpox until 1997.

The Vector smallpox reactor is now reportedly in disrepair. No foreigners were allowed into the space-suit areas of Corpus 6 until 1999, when a team of American scientists went inside. The area had been sterilized, and they didn't wear space suits, but they did wear Level 3 outfits. They noticed the pox bioreactor and asked what it was. A Vector employee replied, with a straight face, in a thick Russian accent, "Is a sewage-treatment fazility."

The Americans were virologists, and they knew exactly what a virus bioreactor was. One of the Americans replied, "Oh, yeah, *right*." The Vector scientists misunderstood the reply and thought the Americans had no problem with their tank. Recently, Sergey Netesov, the deputy director of Vector, insisted in an e-mail to an American government scientist named Alan Zelicoff that, indeed, the Vector pox reactor really is a sewage-treatment tank. "Sergey's lying—he is simply lying," Zelicoff said to me. "I am reminded of how Teddy Roosevelt said that Russians will lie even when it is not in their best interest to do so."

The Vector scientists are dead broke. Some of the Vector weapons-production tanks are now occasionally used to manufacture flavored alcohol, which is marketed in Russia under the brand name Siberian Siren.

No one seems to know what happened to the many tons of frozen smallpox or the biowarheads. Today, both the Zagorsk Virological Center and the bioweapons facility at Pokrov are under extremely tight military security. Both sites are controlled by the Russian Ministry of Defense. They are closed to all outside observers and have never been visited by bioweapons inspectors or by representatives of the WHO. "When we approach people in those places," Alan Zelicoff said, "the door is literally slammed in our faces. We are told to go away. I think the conclusion is that they are going ahead with BW [biowarfare]." The Zagorsk and Pokrov military officials have never offered the world any evidence that the many tons of smallpox once stored at these sites were destroyed. "The sixty-four-thousand-dollar question is what happened to the smallpox material for those warheads," one source close to the situation said. "All we've ever gotten from our Russian colleagues is bland assurances like, 'If it ever existed, it's gone.' It's hard to get them to admit they charged the warheads with smallpox. We don't know where the warheads are now. If they were charged with high-test smallpox, how were they decontaminated? We ask them, 'Did you drain the warheads?' and we don't get an answer. If those warheads weren't drained, then they have smallpox in them now."

Nobody seems to trust the Zagorsk military virologists, not even other Russian bioweaponeers. The Vector scientists have been known to refer to them privately as *svini*—swine. The U.S. State Department circulated an internal brief indicating that Lev Sandakhchiev had been quoted in *Pravda* as saying that he was worried about the "probability that smallpox samples may exist in laboratories other than Novosibirsk [Vector], for example, in Kirov, Yekaterinburg, Sergiyev Posad [Zagorsk], and St. Petersburg." Sandakhchiev later insisted that *Pravda* got him all wrong: "That I never said. This is insane!"

"Lev was no doubt punished for his remarks," Zelicoff observed. "I'll bet my paycheck the Russians have clandestine stocks of smallpox at Zagorsk," another American government scientist who had spent time at Vector said to me. "The Russians themselves have told us that they lost control of their smallpox. They aren't sure where it went, but they think it migrated to North Korea. They haven't said when they lost control of it, but we think it happened around 1991, right when the Soviet Union was busting up." A master-seed strain of smallpox virus could be a freeze-dried bit of variola the size of a toast crumb, or it could be a liquid droplet the size of a teardrop. If a teardrop of India-1 smallpox disappeared from a storage container the size of a gasoline tanker truck, it would not be missed.

A MICROBIOLOGIST named Richard O. Spertzel was the head of the United Nations biological-weapons inspection teams in Iraq—the UNSCOM teams—between 1994 and 1998. Spertzel joined the Army in the late nineteen fifties and was assigned to the American biological-weapons program at Fort Detrick, where he served as a veterinarian and medical officer. When the biowarfare program was shut down in 1969, he stayed on at USAMRIID, working the peaceful side of biodefense. He knows a good deal about biological weapons. Spertzel is now in his late sixties, a stocky man with glasses and a white flattop buzz cut. He has an understated, blunt way of talking. He made some forty trips to Iraq, until the inspectors were kicked out for being too nosy. Spertzel picked his way through suspected sites of biological-weapons research and development, and he directed the analysis and destruction of the main Iraqi anthrax plant, Al Hakm, a complex of buildings on a missile base in the desert west of Baghdad. The UN teams blew up Al Hakm with a large amount of dynamite. Spertzel now lives on a ten-acre spread in the country just outside Frederick, Maryland, within a few minutes' drive of USAMRIID.

"There is no question in my mind that the Iraqis have seed stocks of smallpox," Spertzel said to me.

"Why do you think that?"

"In a nutshell, the Iraqis formally acknowledged to us that they were acquiring weapons of mass destruction by 1974," he said. By then, Spertzel explained, the Iraqis had already built a pair of Biosafety Level 3 lab complexes at a base called Salman Pak, which covers a peninsula that sticks out in a bend of the Tigris River. Salman Pak was run by the Iraqi security service. They had what they called an "antiterrorist training camp" there. "It would have taken a while to build these biocontainment labs at Salman Pak, so we think their biowarfare program dates back to 1973 or earlier," Spertzel said.

In 1972, an outbreak of smallpox occurred in Iran and spread into Iraq. "There would have been many samples of smallpox in hospital labs in Iraq after that outbreak," Spertzel said. "It is inconceivable to me that at just the time when they were starting a biowarfare program they would have gone around Iraq and thrown out all their smallpox."

In the mid-nineties, the UN inspectors often used the Habaniya air base outside Baghdad. Every time they flew into Habaniya and took the road to town, they drove past a group of dusty concrete buildings that were run by a branch of the Ministry of Health called Comodia. The Comodia buildings were warehouses and repair shops, and they were surrounded by apartment buildings and residential neighborhoods. This did not seem to be a likely place for biowarfare activity, but in Iraq you could never be sure, so one day the inspectors decided to have a look around Comodia.

The repair shop was a nothing. They went into the warehouse. On the second floor they found a machine sitting by itself in its own room, awaiting repair. The inspectors recognized the machine as a type of freeze-dryer that is used for filling small tubes with seed stocks of freeze-dried virus. The machine had a label on it that said SMALLPOX.

"I just hoped they'd sterilized the thing," Spertzel remarked.

The top virus expert in the Iraqi biowarfare program was Dr. Hazem Ali, a beefy, robust, proud man in his forties, who had a Ph.D. in virology from Newcastle University in England. He spoke fluent English with a British accent. "He was one of the more brilliant scientists we had contact with," Spertzel said. Dr. Ali ran a complex of Level 3 biocontainment labs called Al Manal, which was Iraq's virus-weapons development facility. Al Manal is in the outer suburbs of Baghdad. The UN people spent some time questioning Dr. Ali in a room in the Al Rashid Hotel, and in September 1995, they questioned him in a conference room where television cameras were operated by the Iraqi government. Spertzel listened while Dr. Ali described his work with poxviruses at Al Manal. Dr. Ali said that he and his group had been working to develop camelpox virus as a biological weapon. Camelpox virus is extremely closely related to smallpox. It makes camels sick, yet it hardly ever infects people—you could run your hands over the wet, crusted muzzle of a pustulated camel, then lick your hands and rub them on your face, and you would probably not catch camelpox.

"You sit back and listen to this, and you try to control your emotions," Spertzel said. "If I heard that from some Joe Blow on the street I would say, 'He's an idiot,' but this was Dr. Hazem Ali, and he is not an idiot, he is a British-educated Ph.D. virologist. Our only explanation for their camelpox was that it was a cover for research on smallpox." The biocontainment zones at Al Manal were kept at Level 3, but the safety controls didn't look like they were up to Western standards. The Americans and most of the Europeans on the UN team were very afraid of Al Manal. They wanted to blow the place up, but the French government vetoed that idea.

Al Manal had been built by a French vaccine company then known as Pasteur Mérieux (now part of Aventis-Pasteur). Pasteur Mérieux had constructed it as a plant for making veterinary vaccines and had run the facility while training Iraqi staff on the equipment. The Pasteur Mérieux people left Al Manal several years before it was converted into a poxvirus-weapons facility, and though they may have been a little naïve, there is no evidence they ever thought Iraq would use the plant for weapons.

In any event, the French government did not want to see a French-built plant dynamited, principally because that might threaten France's other commercial interests in Iraq. The United Nations had to find a less obvious way to give the facility the deep six. "We filled the air-circulation system with a mixture of foam and concrete before we left Iraq, and I believe we made the labs unusable," Spertzel said. Not that it matters. A Level 3 lab is not expensive to build or very difficult to hide. Most legitimate Level 3 research facilities are a few rooms, and they can be anywhere.

In 1999, the Iraqi government asked the United Nations for funds to reopen Al Manal. The UN turned down the request.

"Their biowarfare program continues," Spertzel said, "and the chance the Iraqis are continuing research into smallpox today is high."

· · ·

AFTER THE American-British inspection team visited Vector in 1991 and found evidence that the Vector scientists were doing genetic work with smallpox and were testing the live virus in a chamber for strategic-weapons systems, its findings were classified. The U.S. government decided to work quietly with the new leadership of the Russian Federation to see if the problem could be settled without getting a lot of attention. If the world learned that Russia had a huge biowarfare program, and one that involved genetic engineering, then other countries might be impressed and tempted to get involved with dark biology. One leading expert close to the negotiations between the United States and Russia said that the diplomatic approach failed; the Russians stonewalled the Americans, and the inspections stopped. "The whole thing went into the sand," he said.

"Their BW [biowarfare] program was like an egg," Frank Malinoski (who had been a member of the inspection team) told me. "We saw the white of the egg, but we didn't see the yolk. They hard-boiled the egg, and they took out the yolk and hid it away."

In 1997, the Russian government suddenly announced that the smallpox collection in Moscow had been moved to Vector. The WHO rubberstamped that decision one year later, and Vector became the only official repository of smallpox besides the CDC.

Today, Vector is largely abandoned, and about eighty percent of the buildings there are in ruins or are not being used. Under the Cooperative Threat Reduction Program, the U.S. government has given millions of dollars to the Vector scientists to help them do peaceful research. Visiting American scientists have been told that delegations of biologists or officials from Iran have visited Vector and tried to hire it as a subcontractor to do unspecified research into such viruses as Ebola, Marburg, and perhaps smallpox. In the American intelligence community, Iran is widely believed to have a vigorous and modern biological-weapons program, which it probably established in response to Iraq's biowarfare program.

No outsiders have ever seen the smallpox freezers inside Corpus 6, but there are two of them, an A and a B. The Vector mirrored smallpox is said to contain one hundred and twenty different named samples of variola. Each of them is probably stored in two or more identical master-seed vials. Corpus 6 is surrounded by razor wire and is under military guard, with a security system that was built by the Bechtel Group and paid for by the U.S. government, in the hope of keeping the Vector smallpox from migrating somewhere else.

Battle in Geneva

PETER JAHRLING, the senior scientist at USAMRIID who was called to the office at four in the morning on October 16th, 2001, when the Daschle letter full of anthrax was being analyzed at the Institute, is the codiscoverer and namer of the Ebola Reston virus, the only type of Ebola that has ever been seen in the Western Hemisphere. Ebola is an emerging virus from the rain forests and savannas of Africa that causes people to die with hemorrhages flowing

from the openings of the body. There are now five identified species of Ebola. The hottest of them, Ebola Zaire,
kills up to ninety-five percent of its infected victims, and
there is no cure for it. Jahrling discovered the Ebola
Reston virus in 1989, during an outbreak of Ebola in
Reston, Virginia, a suburb of Washington, D.C. Before he
knew what the virus was, he inadvertently inhaled a whiff
of it from a small flask. Tom Geisbert, the USAMRIID
microscopist whom Jahrling would later ask to examine
the Daschle anthrax, also took a whiff. The two scientists
tested their blood every day for a while after that, but they
never became sick. They are the official codiscoverers of
Ebola Reston, and they have continued to collaborate on
research into Ebola. Peter Jahrling also discovered that an
antiviral drug called ribaviran can be used successfully to
cure people who are infected with Lassa, the Level 4 virus
that turns people into bleeders.

In the nineteen nineties, as the presence of biological
weapons in Russia and other countries became more obvious and more alarming, Peter Jahrling expanded his interests beyond Ebola and began to study smallpox. He
worked with the Cooperative Threat Reduction Program,
and he flew frequently to Vector, where he got to know
Lev Sandakhchiev, Sergey Netesov, and many other Vector
people. He exchanged Christmas cards with them every
year and drank vodka with them when he visited. He liked
them personally and tried to get along with them.

In the late nineties, there was virtually no smallpox vaccine on hand in the United States—at any rate, nowhere
near enough to stop even a small outbreak. Jahrling got
involved in efforts to create a national stockpile, but he
came to believe that the vaccine would not be sufficient if
there was a bioterror attack on the United States. The traditional vaccine, vaccinia, has a rate of bad reactions,
including brain disease and death, that probably makes it

unacceptable by modern standards of pharmaceutical safety. About one in five people couldn't receive the vaccine under current rules. The vaccine is a live virus, and it can sicken or kill people who have lowered immune systems. Today, many people have lowered immunity, including those who are taking immunosuppressive drugs, such as people in chemotherapy or people with inflammatory diseases. Many people also have lowered immunity because they are HIV-positive. The vaccine can't be given to people with eczema, or to family members of someone with eczema or other skin conditions, and it can't be given to pregnant women or to families that have a baby in the house. The pustule formed by a vaccination is contagious if it oozes. If the smallpox vaccine was given indiscriminately to everyone in the United States, it is suspected that at least three hundred people would die, or perhaps one thousand or more—no one really knows—and many other people would be sickened. If a pharmaceutical company marketed a drug that killed a thousand people, it would be one of the biggest scandals in the history of the drug industry.

Peter Jahrling had a loose-knit group of researchers around him, and he pushed them to develop other ways of protecting people against smallpox. He was encouraged by the increasing success of antiviral drugs in fighting HIV.

One of Jahrling's collaborators at USAMRIID, a virologist named John Huggins, ran some experiments and found that a drug called cidofovir (which is marketed under the brand name Vistide) could be used to successfully treat monkeys infected with monkeypox. Working in the Maximum Containment Lab at the CDC in 1995, Huggins also found that cidofovir seemed to work against smallpox in a test tube. Cidofovir might help people with smallpox, and possibly other smallpox drugs could be found. An antiviral drug for smallpox could also be used to treat people who had bad reactions to the existing vaccine; it could be a safety net for immune-compromised people in case millions of people needed to be vaccinated for smallpox quickly.

In order to develop drugs and a new vaccine for smallpox, it would be necessary to do experiments with live variola. The Food and Drug Administration would never license a drug or vaccine for smallpox unless it had been tested and shown to work on at least one type of infected animal. Two centuries ago, Edward Jenner had tested his vaccine in a human challenge trial. Human challenge trials with real smallpox today would be unethical and highly illegal, and might well be considered a crime against humanity. There were going to have to be other ways to test cures for variola besides Edward Jenner's way.

SHORTLY BEFORE the Eradication was formally declared complete in December 1979, D. A. Henderson moved to Baltimore and became the dean of the School of Public Health at Johns Hopkins University. He and his family settled into a solid brick Georgian house near the campus. They built a Japanese garden along the side of the house, and D.A. enjoyed entertaining students and faculty there. He loved to spend a Saturday in the family room, in a big easy chair by the sliding glass doors that looked out on the garden. For years, his wife, Nana, had been asking him if he had any plans to retire; he said he would like to retire, but not immediately. He served as a presidential science adviser in the Bush, Sr., White House for a while, and he has a top secret–level national-security clearance. He began hearing about the Soviet/Russian biowarfare program in the mid-nineties. Starting in 1995, the government gave national-security clearances to people involved with public health, microbiology, and smallpox. Many of them were taken into a conference room at USAMRIID and briefed by Peter Jahrling and others who had special knowledge. They were also briefed by Ken Alibek, the second major defector to come out of Biopreparat after Vladimir Pasechnik.

D. A. Henderson was dismayed by what he learned. He was slow to accept the disturbing information that he was getting about smallpox in the Soviet Union, and he could hardly bear to confront it. Soviet public health doctors had been the early driving conscience behind the Eradication, and the country had donated much vaccine to the effort. Svetlana Marennikova, the keeper of the WHO's smallpox in Moscow, had seemed to be a thoroughly professional scientist. It wounded Henderson to accept this, but by early 1997, he had concluded that smallpox was by no means under control in just two freezers. What shocked him the most was the revelation of the twenty tons of smallpox at Zagorsk. In his mind, this was an obscenity. As early as 1998, he became alarmed about Osama bin Laden, and he began making public statements about the possibility that bin Laden's organization would acquire smallpox. He began working behind the scenes to encourage the U.S. government to build up a stockpile of the smallpox vaccine, but he found this hard going, since no one seemed to take the threat very seriously. Nobody, except for a handful of people like Peter Jahrling, seemed to understand how bad the disease was or how fast it would spread. Increasingly concerned about the threat of biological terrorism, Henderson founded the Johns Hopkins University Center for Civilian Biodefense Strategies and became its first director.

One gray winter day in 1999, I visited Henderson in his house, and we sat in the family room and ate ham sandwiches and drank Molson beers. He was older yet the same man—six feet two, broad shouldered, with a seamed, angular face, pointed ears, and a thick brush of hair, though now gray. He filled the room with his gravelly voice and an aura of human power. He was the medical doctor who had driven variola out of humanity. The walls and shelves of the room were crowded with African and Asian sculptures and wooden Ethiopian crosses that he had picked up during his travels. "If smallpox were to appear anywhere in the world today, the way airplane travel is now, about six weeks would be enough time to seed cases around the world," he said. "Dropping an atomic bomb could cause casualties in a specific area, but dropping smallpox could engulf the world." He sipped his Molson, and the sky turned the color of bluestone, and raindrops splattered across the wooden decks in the Japanese garden.

At that time, very few public health experts or government officials took D. A. Henderson seriously when he said he thought a global smallpox outbreak could really happen. He was viewed in Washington as an older guy who had become a pain in the neck. Henderson intended to remain a pain in the neck for the foreseeable future. He had preserved his top-secret national-security clearance, because he believed that if a bioterror event occurred, the government might want to pull him in to help, and he would need a security clearance in order to serve. Because he had the clearance, he heard about little bioterror threats that didn't get into the news. He felt they were harbingers of something bigger. "In the last ten days," he remarked to me, "we've had fourteen different anthrax scares. Everybody and his brother is threatening to use anthrax. Of course, a real bioterror event is going to happen one of these days."

In a calm, persistent voice, he argued for the destruction

of the official stocks of variola. "What we need to do is create a climate where smallpox is considered too morally reprehensible to be used as a weapon," he said. "It would make the possession of smallpox in a laboratory a crime against humanity. The likelihood that the virus would be used as a weapon is diminished by a global commitment to destroy it. How much it is diminished I don't know. But it adds a level of safety."

HENDERSON was a member of the Ad Hoc Committee on Orthopoxvirus Infections, the smallpox advisory panel for the WHO. The committee was composed largely of veterans of the Eradication, and they met at irregular intervals in Geneva. Starting in 1980, they began to discuss getting rid of the two repositories of the virus—the one at the CDC and the one in Moscow. Henderson now says that at that time he didn't care much one way or the other whether the stocks were destroyed, since the disease had been eradicated, and that was the main thing. Between the CDC freezer and the Russian freezer, there were not more than a couple of pounds of frozen smallpox material, all told. The vials would have fit into a few cardboard boxes, and heating them in an oven would seem to be easy.

Some of the committee members felt that destroying the stocks of smallpox in Atlanta and Moscow amounted to the purposeful extinction of a species. Even though it was variola, the worst human disease, would it be proper to send it to extinction? (They didn't know that the Soviet Union was then making variola by the ton for loading into ICBMs.)

In 1990, the U.S. secretary of health, Louis Sullivan, asked the WHO what its position was: should smallpox be made extinct as a species? The Ad Hoc Committee solicited the views of the major societies of microbiology, along with the Russian academy of medical sciences. The answers came back, and they were unanimous: variola should die. Nobody wanted variola kept around. Even so, the committee proposed that the information in the DNA of smallpox be preserved. In 1991, the CDC pox virologist Joseph Esposito and the genomic scientist J. Craig Venter decoded the entire DNA of the Rahima strain of smallpox. The genetic information in the Rahima strain could be kept, while Rahima and its fellow strains could be made extinct.

In 1994, the committee and the World Health Assembly voted unanimously to destroy all the stocks of smallpox, and they set a deadline of June 30th, 1995, for the execution. The official stocks would be cooked in autoclaves—ovens that would make the vials of smallpox sterile. But then, abruptly, the British Ministry of Defence and the U.S. Department of Defense began to object to the plan. The 1995 deadline passed, and the stocks of smallpox still sat in the freezers.

The governments of nonindustrial countries that had suffered from smallpox didn't like the idea of American and British military people keeping smallpox: it made them very nervous. In 1996, the WHO General Assembly voted for the total destruction of the official stocks and set a new deadline of June 30th, 1999, but as that deadline approached, opposition to the destruction persisted. It was now coming both from Russia and from some members of the American scientific community, mainly virologists, who wanted to study smallpox for pure scientific curiosity. In the summer of 1998, the Institute of Medicine, a branch of the National Academy of Sciences, formed a panel of experts to explore the question of what sorts of important research might require real variola. D. A. Henderson smoldered about it. He objected to the very way in which the panel was posing the question: "If you ask a scientist what research could be done if he had the live smallpox virus, of *course* he's going to tell you a lot of research could be done." In his view, there was no good scientific justification for research into real variola. He had no illusions any longer that variola sat in only two freezers, but he felt that the United States and Russia had an opportunity to show the world the moral high ground. He felt that the traditional vaccine had worked in the Eradication and would work again if there ever was a bioterror attack with smallpox. He thought that an antiviral drug for smallpox was a long shot that would waste resources, and that the research would interfere with the far more important task of showing the world that the United States and Russia could get along fine without smallpox. "To get a new drug for smallpox will cost three hundred million dollars, and the money simply isn't there," he said.

ON January 14th, 1999, the Ad Hoc Committee on Orthopoxvirus Infections met at the WHO, in a conference room in an annex building. The meeting was chaired by D. A. Henderson. The participants were the inner circle of the committee. There were also a number of straphangers—people sitting in chairs at the edges of the room, sometimes asking questions. One of them was Peter Jahrling. Lev Sandakhchiev, the head of Vector was in the inner circle of the eradicators. Sandakhchiev is a chain-smoker, and at every break he went outdoors and paced along a walkway in the cold, wreathed in a cloud of pungent blue smoke from his Russian cigarettes.

Lev gave a presentation. He read from a long, prepared

text in English, and when he answered questions, Svetlana Marennikova, the former guardian of the WHO smallpox repository in Moscow, helped to translate for him into English. Sandakhchiev said that there had been no work with smallpox at Vector until very recently. Though the WHO smallpox had been moved to Vector in 1994, he said, it had sat in a freezer for three years, and nobody there had used the virus for experiments until 1997.

This caused a big stir in the room: the WHO had not officially made Vector the repository until 1998, but here Sandakhchiev was saying that the smallpox had been moved out of Moscow without anyone's permission and years earlier than had been supposed.

The Japanese eradicator, Dr. Isao Arita, was particularly dismayed, and he grilled the Russian pox scientists: "Why did you move it then? Why didn't you tell us?"

Henderson had good reason to believe that Sandakhchiev and his people had been developing and testing smallpox for weapons in 1990. "I rolled my eyes," he recalled later. "And I saw Peter Jahrling and other people rolling their eyes at me. It was quite elaborate and quite unbelievable. We're sitting there, he's presenting us with all this horseshit, and he knows it's horseshit. He was lying flagrantly."

Toward the end of the meeting, Henderson gave his views. He spoke for forty-five minutes in his gravelly voice, with passion and restrained anger. He said that Osama bin Laden represented a danger to the world. He said that bin Laden could get smallpox and would use it. He said that the Aum Shinrikyo sect in Japan could get smallpox and use it. He said if smallpox was used as a bioterror weapon, everyone on earth was in danger, and it was imperative that the leading countries of the world agree to destroy the official stocks of variola. Looking straight at Sandakhchiev and at his old colleague Marennikova, he said he believed that smallpox existed in at least *three* places in Russia. He said that several biowarfare scientists from Russia had "gone south"—to countries in the Middle East. He spoke of his opposition to any further research with variola in the laboratory, and he posed a question to the room: "How many requests for experiments with variola have you actually had in the past twenty years?"

Lev Sandakhchiev firmly insisted that he and his people hadn't been doing anything with smallpox until very recently.

As for the CDC, it had had virtually no requests for experiments with smallpox. The smallpox had been sleeping peacefully in the freezer, except when the Rahima strain was taken out and its DNA was sequenced by Esposito and Venter. But Peter Jahrling wanted to awaken the smallpox stocks and use them in research. Speaking from his seat as an onlooker, he said, "D.A., the tools of molecular biology have advanced quite a bit in the past twenty years. Just because there hasn't been any demand for variola in the past, it doesn't mean there won't be demand for variola in the future."

Henderson answered by saying that the arguments for keeping smallpox in order to do research for antiviral drugs or better vaccines were completely specious. He said that a new vaccine would require an animal model, and *that* we would never have. Smallpox would not infect an animal; it was a virus of people. It is safe to say that D. A. Henderson was in almost unbearable agony as he gave that speech.

The meeting ended with a vote on whether to keep or destroy the stocks. It went five to four in favor of destroying smallpox—a narrow victory for Henderson. But the virus had already passed out of the control of the WHO, and Lev Sandakhchiev had more or less said so. Henderson felt terrible regret that he and the other members of the committee hadn't voted to destroy the stocks in 1980, right after the Eradication. Everyone would have agreed to it, and they could have done it then.

THE EXECUTION DATE of smallpox, June 30th, 1999, began to loom. In April, the Institute of Medicine issued a report saying that if the world wanted to have a new vaccine or an antiviral drug for smallpox, then the virus would need to be kept for scientific experiments. President Bill Clinton had personally favored the destruction of smallpox, but the report changed his mind, and the White House now strongly endorsed the idea of keeping the stocks. A month later, the WHO General Assembly voted to keep smallpox alive for another three years, until June 30th, 2002. Researchers—principally Peter Jahrling and his group—could use that time to see if it would be possible to cure smallpox with a drug or if it would be possible to find an animal that could be infected with smallpox so that new vaccines could be tested.

The stakes could not have been higher for Jahrling. He felt that a smallpox emergency could be as bad as a nuclear emergency. "Smallpox is the one virus that can basically bring the world to its knees. And the likelihood of smallpox being visited on us is far greater than a nuclear war, in my opinion," he said to me. Now he had three years to do something about it. There were times when he woke up, bolt upright in bed, sleepless at three o'clock in the morning with *Pocken-angst*, with anxiety about smallpox. He would talk to D. A. Henderson in his head: *Goddamn it, D.A. . . .* He and the eradicator agreed perfectly on the nature of variola—it was the mother of all biological weapons. But they could not agree on what to do about it.

A Woman
with a
Peaceful Life

Lisa Hensley

ON THE FIRST of September 1998, a twenty-six-year-old civilian scientist named Lisa Hensley reported to work for the first time at USAMRIID. She was a postdoctoral researcher with a fellowship from the National Research Council. Hensley rented a one-bedroom apartment in Germantown, about twenty minutes outside Frederick. She furnished the apartment with a couch and a television set that she had inherited from her grandmother.

Lisa Hensley is of medium height, with hazel eyes and dark brown hair that she usually wears tied back in a ponytail; when she's working in the laboratory, she ties her hair up so that it won't fall into her experiments. She has an open face, a calm, unruffled manner, and a rapid, precise way of speaking. She was an All-American varsity-lacrosse player at Johns Hopkins, and she has broad shoulders and an athletic way of moving. She usually wears khaki slacks, square-toed loafers, and gold earrings decorated with small pearls. She rarely takes off the earrings, even when she's inside a biohazard space suit. Hensley is a scuba diver, and she likes to dive on wrecks and into underwater caves. Cave diving is not for people who get claustrophobia, and the sport has a high rate of accidents. She finds it calming, she says.

Lisa's father, Dr. Michael Hensley, works in the pharmaceutical industry. When he was younger, Mike Hensley rode horseback and fenced with sabers, but during his medical internship he had what he describes as an interesting event—a hemorrhage. He learned that he had a mild form of hemophilia, a genetic disease that occurs only in men but is inherited through female carriers. Hemophilia ran in the Hensley family. Many men with hemophilia have died of AIDS, having received blood transfusions tainted with HIV during the years when blood wasn't tested for it.

When Lisa was eight years old, HIV was just beginning to be understood. Mike Hensley received blood transfusions during that time, but he didn't become infected. Lisa was extremely close to her father. He took her into his laboratory and taught her how to grow bacteria on petri dishes, and he gave her bottles of seawater to look at in his microscopes. She saw that a tiny droplet of the sea was an ecosystem packed with life. She told her parents that she wanted to be a marine biologist, and at twelve she was certified as a diver.

In high school, she was a jock who was bored out of her mind with her studies, including biology. She became a state-champion lacrosse goalie with a string of varsity letters, and applied to the U.S. Naval Academy to become an aviator. Then, at the last minute, she changed her mind and went to Johns Hopkins, which recruited her to play lacrosse.

At Johns Hopkins, Hensley began taking courses in public health. When she was a junior, Mike Hensley invited her to attend a scientific conference in San Francisco on HIV, and it electrified her. She became fascinated with the idea that if you really understood how viruses emerge, you might be able to stop a disease like AIDS before it could spread. She graduated from Johns Hopkins in four years with a master's degree in public health.

Hensley went on to get a Ph.D. in epidemiology and microbiology in three years at the University of North Carolina at Chapel Hill, and at the same time she got a second master's degree in public health. She had pretty much no social life in graduate school and devoted herself to the laboratory. She moved viruses from one type of host to another and watched trans-species jumps occur in the lab, before her eyes. She learned the standard methods of virus engineering—how to change the genes of a virus, altering the strain.

Hensley had an apartment across the street from her lab at Chapel Hill, so that she could spend nearly every waking minute in the lab, with the goal of having three advanced degrees by her twenty-fifth birthday. She didn't sleep much, and when she did she had recurrent dreams, focused on her hands. In the dreams, she was working faster and faster, trying to finish an experiment, yet she could never make her hands go fast enough. . . . She was falling behind. . . . Her grant money was running out. . . . Life was too short. . . . And she would wake up. She'd grab a Diet Coke for breakfast and stumble across the street to the lab, where she would work all day and half the night.

At USAMRIID, Lisa Hensley began doing research on SHF, a Level 3 virus that is harmless to humans but is devastating to monkeys. It was a virus that could emerge as a human disease someday. Her social life had opened up, and she had begun dating a virologist at the National Institutes of Health in Bethesda, Maryland. Things didn't work out well between them. The problem was that when they argued with each other, it was about viruses. Scientific people are competitive types, and they like to be right. Any sort of discussion about viruses with her friend could turn into an emotional fight. One time, they were in his apartment debating some minor point about a virus, and he said, "You're wrong about that." She went over to a shelf, grabbed a textbook, and opened it to the page that showed she was right. She placed it on the kitchen table and walked out. Hensley admitted to herself that this was perhaps not emotionally shrewd. When they broke up, she vowed to herself, *No more scientists, they're a headache.*

The head of Lisa Hensley's division at USAMRIID was Colonel Nancy Jaax, an experienced pathologist with a strong interest in Ebola virus. Hensley had zero interest in Ebola. The space suits at USAMRIID are blue, and from the day she arrived there Hensley made a point of saying, "The people who work in the blue suits are nuts. I'm not putting on a blue suit for Ebola. You have to be crazy to do that."

Nancy Jaax heard about Hensley's cracks about people who worked with Ebola being crazy. It was felt that cautious people would be less likely to have an accident in Level 4. The last thing anyone wanted was a researcher getting cocky around a hot agent.

One day Hensley walked into a regular staff meeting, in a windowless conference room on the second floor of the Institute, and, as a junior scientist, took her place at the foot of the table. The meeting droned on for a while, typically, until Nancy Jaax suddenly looked down the table at Hensley and announced that her mission was about to change. "I'm going to have you refocus your efforts, Lisa," Jaax said. "We'll get you trained in the blue suits, and we'll start you working with Ebola Zaire."

Lisa Hensley came out of the meeting feeling dizzy and a little unsteady on her feet. She teetered back to her cubicle and fell into a chair. The cubicle was a cluttered space, piled with papers. There was a computer, a stereo, and pictures of her mother and father and other members of her family. *They're going to start me with Ebola Zaire?* she thought.

Death from Ebola comes about five to nine days after

you break out with symptoms, and it occurs with spurts of blood coming from the orifices and a collapse of blood pressure, an event that Army people call the crash and bleed-out. In some cases, the virus causes a near-total loss of blood—an Ebola exsanguination. They were paying Hensley thirty-eight thousand dollars a year, but was it worth it? If you infected yourself with Ebola, that was it.

Hensley was closer to her parents than to anyone else in the world. Her mother, Karen, called her three times a week to find out how things were going. Lisa told her that the powers at the Institute had redirected her career into Ebola virus.

"You're going into a BL-4 suite to work with *Ebola*? Isn't there anything else they could have you do?"

Lisa tried to play things down. "Oh, Mother. I'm much safer in a space suit. Really."

"Mike! Mike! Come talk to your daughter."

Lisa's father thought it was a good opportunity for her, and they decided to give Karen a tour of the laboratories, so she could see that everything was safe.

The tour wasn't quite as successful as they had hoped. Karen Hensley is an economist, and she didn't have a natural feel for biohazard containment. She noticed a door marked CRASH DOOR. That didn't sound too good, but it was a safety feature. If a fire or other emergency occurs in Level 4, you can burst out through the crash door, and you end up standing in the hallway in your space suit. (So far, no crash door has been used for that purpose at USAMRIID.) What really bothered her was the fact that the edges of the crash door were sealed with brown duct tape, which ran all around the door frame. "Why do they have tape around that door, Lisa? Is that how they seal the doors around here? Just with tape?"

Lisa explained to her mother that the hot suites were under negative air pressure, and air was constantly flowing *into* the labs, so the tape was actually to prevent dust and contaminants from entering and messing up the experiments.

Karen Hensley didn't like the look of the tape, period. Then she discovered what you wear inside a biohazard space suit: green cotton surgical scrubs, latex surgical gloves, and socks. That is all. Underwear is forbidden in a hot lab. Karen Hensley was mortified for her daughter. She could not imagine why they would make any woman work in a laboratory without a bra.

HENSLEY WAS TRAINED in blue-suit work by an older postdoc at USAMRIID named Steven J. Hatfill, a big, muscular man in his forties with a mustache and a medical degree—a civilian medical doctor with a background in the U.S. Special Forces. He showed her how to put on the suit, how to do a safety check on it for leaks, how to maintain it properly, and how to go in and out through the decon-shower air lock in Level 4. Steve Hatfill was known around the Institute as a "blue-suit cowboy." He seemed fearless in a blue suit, and he thrived in Level 4. He had a thirst for adventure: he had been a soldier in Africa, where he said he had served in Rhodesia with the white Rhodesian Special Air Squadron—the SAS—during the years when black insurgents were trying to overthrow Rhodesia's white government. Later, he got a medical degree in Zimbabwe, and he worked as a doctor in Antarctica for a year and a half with a team of South African scientists. Hatfill had become convinced that a bioterror event was likely to happen. He served as a consultant to emergency planners in New York City, and he kept a strip of reflective tape on the roof of his car, so that in the event of a bio-emergency the state-police helicopters could find him.

Lisa Hensley found Steve Hatfill likable and entertaining, quite a character. He was bright, a super lab worker, and he taught her some techniques. He was researching the coagulation of monkey blood infected with Ebola virus. Ebola blood became hemorrhagic and wouldn't coagulate, but it needed to be clotted in the lab for study; he taught her how to do this. He had all kinds of gadgets running in Level 4—assay machines, that sort of thing.

During one of her first training sessions, Hensley looked over at Hatfill and noticed that he was hunched inside his space suit. One arm of his suit was hanging limp, as if he had had a stroke. At first, she didn't know what was going on: was he suffocating or what? Hatfill had pulled his arm up inside the sleeve of his space suit, and he was eating a candy bar.

LISA HENSLEY was a rising star at the Institute. Postdocs like her tended to move on quickly if they got bored, and she was assigned to work in Peter Jahrling's group. Despite his growing involvement with smallpox and national policy, Jahrling had continued to do research into Ebola virus, working closely with Tom Geisbert. They not only were scientific collaborators but had become personal friends. Lisa Hensley went to work for Geisbert, who was running Ebola experiments in Level 4. She did lab work on samples of monkey blood infected with Ebola. On her own, she began developing tests for detecting the presence of Ebola virus inside individual cells. The tests made infected cells glow red or green under a fluorescent microscope. You could see how Ebola virus was invading cells in the immune system and doing clever things that seemed to trigger a cytokine storm. She was getting closer to understanding how Ebola

overwhelms the human immune system. This was important work, because there might never be a vaccine or cure
for Ebola unless scientists understood how it killed.

Hensley found that she liked the peaceful feeling of
working alone in a space suit in Level 4, with nobody to dis-
tract her, nothing but the green cinder-block walls and her
dishes of Ebola. She felt cozy inside the suit, even though
the rooms around her were hot with the virus. It was like
scuba diving. A space suit was a sanctuary from the hubbub
of the world. You could do your work without being interrupted by people asking questions or calling on the telephone, and you could press a little deeper into nature.

Hensley was growing Ebola in virus cultures. Viruses
are grown in plastic well plates containing a liquid cell-
culture medium. In the bottom of the wells there is a carpet of living human cells, alive and bathed in the liquid.
(The cells are HeLa cells, cervical-cancer cells derived
from an African-American woman named Henrietta
Lacks, who died in Baltimore in 1951. Her cancer cells
have become a cornerstone of medical research and have
saved many human lives.) Hensley would infect plates of
cells with Ebola, and in a few days virus particles would
begin budding out of them. Ebola particles are shaped like
spaghetti, and they grow out of the cells like hair. The
strands break off and drift away in the liquid. The virus is
amplified in the well plate, and in a few days the liquid
becomes a virus soup, rich with particles of Ebola.

Hensley became good at making amplified Ebola
soups. Using a pipette, she moved droplets of Ebola soup
around from well to well, from vial to vial. She would
hold the pipette in her heavy yellow rubber gloves, push
a button on the pipette with her thumb, pick up a small
quantity of the Ebola soup, and then drop it into a vial.

Ebola soups are pale red, the color of a watered ruby,
and sparkling clear. A well plate full of Ebola soup con-

tains up to five million lethal doses of the virus—in theory, enough Ebola to make half of New York City crash and bleed out. Yet handling Ebola soup is no more dangerous than walking down a busy street. You could be killed if you stepped in front of a bus, but careful people watch where they are going. Hensley wore earplugs, and she heard nothing but the distant roar of cool, sterile air running in her space suit. It sounded like surf on a beach.

Hensley spent so many hours working in her suit with Ebola that she began to get those dreams again. In her Ebola dreams, she would be moving droplets of Ebola soup from well to well, from vial to vial, working faster and faster, trying to complete an experiment, and there was never enough time to find out what she longed to know about viruses. In her dreams, she was always in control of Ebola virus, and Ebola never had control over her.

Panic in the Gray Zone

JANUARY 12, 2000

HENSLEY WAS WORKING ALONE in her blue suit in hot suite AA4, near the center of the Institute. It was about three o'clock in the afternoon. She had been working with soups of Ebola for many hours. She wasn't feeling well: she had a cold and was a little achy, as if she might have a slight fever. She had probably caught a virus, but she was in the middle of an experiment, and she couldn't abandon it just because she felt sick. She would lose her data if she went home.

She was holding a pair of blunt children's scissors with her rubber space-suit gloves. (Sharp scissors are forbidden in Level 4.) She was trying to open a bottle by prying on a tab with the scissors. Suddenly, they slipped, and the tip of the scissors jammed into the middle finger of her right glove. She felt a stab of pain near her fingernail.

She held her space-suit glove in front of her faceplate. What had just happened? Had she cut the glove? The yellow rubber was wet, and as she turned the glove in the light, she couldn't tell if there was a cut in the rubber or not.

Inside her space-suit gloves, she was wearing latex surgical gloves, for an extra layer of protection. Wriggling her arm, she pulled her hand out of the space-suit sleeve and up inside her space suit—the way Dr. Hatfill did when he ate a snack inside his suit—and inspected the latex glove at close range, inches from her eyes.

The rubber was translucent. Beneath it, she saw blood oozing out of her finger, moving along the fingernail. A red spot under the rubber was spreading along the cuticle. It hurt.

It is believed that a single particle of Ebola virus introduced into the bloodstream is fatal.

Hensley felt a sudden rise of fear, which turned into a little bit of panic. *What was the last thing I touched with my hand? What was I doing? What were the scissors touching? Was there any soup on the scissors?* The mind goes sticky in a moment of fear. She blanked. She couldn't remember what she had been doing with her hands. There was nobody to ask.

She began to talk to herself silently: *Quit panicking*

and calm down. Did I make holes in both gloves? Or did I just crush my cuticle? She stuffed her arm back down into the sleeve of the space suit and wiggled her fingers into the outer glove.

Time to get out of here.

She opened the decon-shower air-lock door, stepped into the air lock, closed the door, and latched it. She pulled the shower chain, and a spray of chemicals hissed down over her space suit. While she was taking the chemical shower, she realized that she did seem to have a low-grade fever. *Oh, this is great,* she thought. *I already don't feel well, and now they're going to take my temperature and then lock me up.* She racked her brain trying to remember what she had been doing with her hand. Her glove had been slippery and wet . . . wet with detergent. Some detergents kill Ebola particles. So if there had been Ebola on her glove, maybe the detergent neutralized it. The shower stopped, and she opened the exit door and went out into a Level 3 staging area and pulled off her space suit.

The staging area is a gray room—in between the hot side and the cold side. It has lots of equipment in it, and along one wall there is a row of hooks that hold blue suits belonging to all the scientists who work in the suite. A laboratory technician, Joan Geisbert, was working in the gray area. Geisbert is a slender, quiet woman with dark, wavy hair, dark eyes, and a serious, intelligent manner. She is married to Tom Geisbert, Hensley's boss. Joan Geisbert is an expert in Level 4 laboratory work, with many years of experience. Hensley trusted her, but she thought she'd better not say anything.

This is no big deal, she told herself. She pulled off her surgical glove and washed her hands with disinfectant soap.

She needed to know if there was a hole in her inner latex glove. That was a serious question. If there was a hole in the glove and there was bleeding on her hand, then there was a chance the scissors had cut her skin. The tip of the scissors could have been contaminated with Ebola virus. Washing her hands would do no good if a particle or two of Ebola had made it into her bloodstream.

Joan Geisbert was puttering with something, not paying any attention to her.

The way to test a surgical glove for a hole is to hold it under a faucet and fill it with water, like a water balloon. If there's a hole, a thread of water will squirt out. Hensley went to a sink, filled her glove with water, and held it up to see. There was nothing, no leaks . . . but when she squeezed the glove, drops of water oozed out of a hole in the finger.

Okay. She had cut her finger in the presence of Ebola Zaire.

"Hey, Joan? I think I screwed up."

"Let me see." Joan Geisbert came over to the sink and inspected her finger and the hole in the glove while Hensley told her what had happened. Geisbert glanced at her with a look of alarm.

Oh my gosh, Hensley thought.

"Get dressed and report to Ward 200. I'll call Tom and have him meet you there."

Ward 200 contains a Level 4 biocontainment hospital suite known as the Slammer. Someone who has been exposed to a hot agent can live there for weeks in isolation, and if they become sick they are tended by nurses and doctors wearing space suits.

Hensley took a water shower, got dressed in civilian clothes, and reported to Ward 200. By the time she arrived, Tom Geisbert was waiting for her. He was flushed and nervous. Peter Jahrling had been paged, and he had broken away from a meeting in Washington and was driving back to the Institute as fast as he could. The ward started to swarm with medical doctors, Army officers, nurses, soldiers, and lab techs. An accident investigation team formed up and examined the cut on her finger. They wanted to know what she had been doing with her hands just before she had cut her glove. They took her temperature and discovered she had a fever. She explained that she thought it was just a cold. They stuck needles in her arm and drew out many tubes of blood. She was too nervous to sit on the exam table, so she leaned against it, and then she couldn't stop pacing around the room.

Tom Geisbert took an Army major named John Nerges aside. "Can you stay with her?" he said. "Talk to her and keep her mind off it." John Nerges is a large, kindly man, and he was concerned about Hensley, but he joked around and kept up a patter with her.

Meanwhile, the investigation team took her latex glove into another room and studied the hole. They measured the distance between the hole and the location of her cut. Maybe the scissors had not made the hole. They held a meeting out of her hearing. She paced up and down the hallway, with Major Nerges at her side. "Can I get you a soda or anything?" he asked.

"Yes, please."

She could see the Slammer every time she passed the doorway. There was a bed with a bio-isolation tent around it, and a dummy lying on the bed. Teams of soldiers at USAMRIID used the dummy to practice handling a contagious patient. Major Nerges came back with a Diet Pepsi. "It's no big deal," he said. She popped the can and noticed that a soldier had walked into the Slammer and had opened the bio-isolation tent.

He picked up the dummy carried it away on his shoulder.

Hensley turned to Major Nerges. "If it's no big deal, why is he taking the dummy away?"

Major Nerges walked over to the grunt and swatted him lightly across the back of the head. "You idiot, she's standing right there," he growled.

• • •

AFTER INTERVIEWING HER for two hours, and studying the glove and her hand, the accident investigators came to the conclusion that there was a low probability that Lisa Hensley had been infected with Ebola virus. Her glove had been wet with detergent, and they felt it would most likely have killed any virus particles. She was free to go home and get some rest. However, she would need to report to an Army doctor twice a day for the next three weeks.

She wasn't sure her mother should know about this. She went to a telephone near the Slammer and dialed her parents' house in Chapel Hill. Unfortunately, her mother answered.

"Hi, Mom. I need to talk with Dad about science."

In a moment, her father came to the phone, and Lisa told him what had happened.

He spent a long time calming her down. "Let's not tell your mother. I'll call you every day." He was worried about her, but he said, "I think you need to suck it up and get back in there and finish your experiment."

"I know. I know I do."

Otherwise, she might never go back.

At six o'clock, after most people had gone home, she returned to the locker room in suite AA4, put on surgical scrubs and her space suit, and faced the steel door that led inward to Biosafety Level 4. It was a matter of turning the handle and pulling the door open. That was not difficult. The staging room was quiet, empty, with only the sound of her breath running inside her faceplate, which was starting to fog up. Through her visor she saw the red, spiky biohazard symbol on the steel door. Suck it up and turn the handle.

Peter Jahrling arrived at the Institute and went looking

for Lisa Hensley. He didn't find her in her cubicle, so he went to Tom Geisbert's office: "Jesus, Tom, she'll never want to go back to the lab. Was she crying? Where is she?"

"She went into AA4, Pete."

"You're kidding."

The door was heavy, and it swung open slowly. She latched it behind her and stepped through the air lock to the hot side.

Up in Geisbert's office, Jahrling was saying, "Would you and Joan mind giving her the talk?"

"What talk, Pete?"

"The one I don't want to give her. About not sharing body fluids with anyone during the incubation period of Ebola."

An hour later, Hensley emerged from the hot suite, chilled and shivery, feeling a little feverish and perhaps a little trembly, but she had finished the experiment.

Joan and Tom Geisbert were waiting for her on the cold side. They invited her out to dinner at a Mexican restaurant, where they bought her some dinner and two beers. The beers helped. Tom and Joan looked at each other, and Tom said to Lisa, "I'm supposed to give you the talk about not sharing body fluids with anyone for a while."

"Yeah?"

"That was the talk."

"You mean when I kiss a guy, no swappin' spit?"

Tom turned red, and Joan burst out laughing.

She assured them it wasn't an issue right then. In fact, Hensley did have a date lined up that night, a first date with a man she didn't know very well. Finally, she phoned him and asked if he wouldn't mind putting off the date, since she had just had a potential exposure to Ebola virus.

He was very understanding.

She drove back to her apartment, which seemed

freezing cold, and so she placed a space heater on the floor by her grandmother's couch, turned it on, lay down on the couch, and wrapped herself in a blanket. Her cat, Addy, curled up with her. Then she began calling her closest friends, and she stayed up late that night on the couch talking with them. She slept for a while, and awoke from a turmoil of blue-suit nightmares. She was boiling hot, her throat was parched, she had a fever. . . . *Where am I?*—and there was Addy, purring at her side.

HENSLEY CONTINUED her work with Ebola virus, somewhat oblivious to the heated arguments between smallpox experts about whether variola should live or die. The smallpox virus was a relic of the past to her, a virus with a seventies feel, like an album by Debbie Boone. She was more interested in trans-species jumps of viruses that were emerging right now.

She also had her mind on children: what was a woman scientist on a fast track supposed to do about children? She started playing volleyball in a league and met a man named Rob Tealle, who became her significant other. He was a builder and general contractor who worked around Frederick—a smart man, but not a scientist. Hensley packed her grandmother's couch into a U-Haul trailer and moved to an apartment in Frederick. She and Tealle became very close. It was the biggest relief in the world to go home after a day in a blue suit and talk about normal things with a normal guy.

A Failure in Atlanta

AFTER THE WHO committee opened a three-year window in which live smallpox could be worked on, Peter Jahrling and John Huggins put together a plan to try to infect monkeys with the virus. The Food and Drug Administration has long insisted that new drugs for a human disease be tested in humans before they are licensed for use. This is not possible in the case of smallpox. Since smallpox has been eradicated, no one is infected with it, and you can't (legally) infect people with a lethal disease just to study them. So the FDA was in a bind over smallpox. It published a draft of a new rule, the Animal Efficacy Rule, which says that for an exotic threat such as smallpox, the FDA would license a new drug or vaccine if it could be tested in *two* different animals that had the disease, and if the disease resembled the human disease—if there was an animal model of the disease.

Peter Jahrling wanted to get an animal model of smallpox that they could test drugs on and that the FDA would accept. Since there is no smallpox at USAMRIID, Jahrling assembled a research team and flew them to Atlanta. He got permission from CDC officials to bring the smallpox freezer out from its hiding place and allow his team to take out the smallpox, warm it up, and try to infect monkeys with it. Jahrling decided to infect the monkeys by having them breathe smallpox in the air, to mimic the way it spreads among humans.

The USAMRIID scientists built a portable aerosol chamber that they called the Monkey Cabinet. It is a huge device made of plastic and steel, and it has wheels, so it rolls. They trucked the Monkey Cabinet and a number of monkeys down to Atlanta and installed them in the CDC's Maximum Containment Lab. Jahrling and Huggins exposed the monkeys to around two million human infectious doses of smallpox. Then Jahrling went back to Fort Detrick to attend to other business, while John Huggins remained in Atlanta, monitoring the monkeys. A few days after they'd breathed enough smallpox to take out a city, some of the monkeys got a flush across their chests, and a couple of them developed a few pimples. After a day or so, the monkeys recovered.

As the experiment was winding down, Jahrling began to feel desperate. He was afraid that D. A. Henderson would label this experiment a failure and would say "I told you so" and that it confirmed the widely held belief that animals could never be successfully infected with human smallpox. The clock was ticking. A year had passed since the WHO had extended the deadline for destroying the smallpox virus, and Jahrling needed data that looked at least vaguely promising, or the WHO might not allow him to do another experiment.

He needed someone to fly down to Atlanta, take some blood from the monkeys, and do a quick study of it right there. Maybe that would show something. He asked Joan Geisbert, but her son was graduating from high school in Frederick, and she was going to be there no matter what. Lisa Hensley could probably run the tests, but she was wrapped up in Ebola research and pretty clearly did not want to be involved with smallpox. He asked her anyway.

"Yeah, I'll do it, sir," she said.

Jahrling began to wonder about this. If Hensley went down to Atlanta and started working with smallpox, what would happen if she enjoyed it? What if people at the CDC were impressed with her? He told Tom Geisbert, confidentially, that he was afraid the CDC people might try to poach her. USAMRIID and the CDC had a history of strained relations. Jahrling told Hensley that he would accompany her to Atlanta.

She was annoyed, and when she was annoyed she tended to rely on Tom Geisbert for advice. She dropped by his office and asked, "Does he think I need a baby-sitter in Atlanta?" Geisbert explained Jahrling's worry about CDC poaching.

Jahrling and Hensley flew to Atlanta at the beginning of May 2000. They sat next to each other on the government-budget AirTran flight, and Hensley was uncomfortable, not to say tongue-tied. At the CDC, they put on blue suits and entered the Maximum Containment Lab. Hensley worked all day taking blood samples from monkeys that had breathed ten million human doses of smallpox and seemed fine.

Three days later, Hensley was back at Fort Detrick with the raw data from her tests. A month later, Jahrling flew to Geneva and presented Hensley's data to the Ad Hoc Committee on Orthopoxvirus. He argued that the data was "suggestive," which meant that the experiment had bombed. D. A. Henderson argued that Jahrling would never be able to infect monkeys with smallpox, that it just wasn't going to work. Jahrling pleaded for another chance, and the committee agreed to let him try again—in another year.

Then, out of nowhere, came a discovery that shook the smallpox experts to their cores.

Nuclear Pox

A FEW MONTHS after the failure of his monkey-model experiment, on a hot Saturday in early September 2000, Peter Jahrling flew to Montpellier, France, for the thirteenth International Poxvirus Symposium. It was held at Le Corum, a modern conference center in the middle of town. The place was jammed with more than six hundred poxvirus experts from around the world, many of them milling in the lobbies and chain-smoking cigarettes. On Sunday afternoon, Jahrling wandered around a lobby where scientists were showing poster papers.

A poster paper is much like the reports that children give in school. It's usually about an experiment that doesn't warrant a full presentation. A scientist makes up a poster that summarizes the experiment, hangs it up and stands next to it, and answers questions.

There were fifty or sixty poster papers hanging on bulletin boards. Jahrling bumped into Richard Moyer, an American poxvirus expert who is the chairman of the Department of Molecular Genetics at the University of Florida in Gainesville. There was a lot of noise and cigarette smoke, and they wanted to talk, so they found a poster that nobody was looking at. Jahrling and Moyer placed themselves to one side, so they wouldn't disturb the scientist standing next to the poster, and they chatted about some of the things they'd been learning. Moyer glanced over at the poster. He stopped talking.

The experiment described on the poster had been carried out by a group of Australian government researchers from the Co-operative Research Centre for the Biological Control of Pest Animals, in Canberra. They were using viruses to try to cut down populations of mice. The scientist who led the work, Ronald J. Jackson, was the man standing beside the poster. Jackson is tall, with a roundish face, dark, short hair, and a nut-brown tan. He was a pleasant-looking man, wearing a yellow short-sleeved shirt and brown pants.

The Australian group had been working with the mousepox virus, which is closely related to smallpox. Mousepox, which is also called ectromelia, cannot infect humans and doesn't make them sick, but it is lethal in some types of mice. The Australian group had been infecting mice with an engineered mousepox virus that was supposed to make the mice sterile. But the engineered mousepox had wiped out the mice.

The mice were naturally resistant to mousepox, and some of them had also been vaccinated. Even so, the engineered virus had sacked them. It had wiped out a hundred percent of the naturally resistant mice and sixty percent of the immunized mice.

The Australian scientists had added a single foreign gene, the mouse IL-4 gene, to natural mousepox virus. The mouse IL-4 gene produces a protein called interleukin-4, a cytokine that acts as a signal in the immune system. By putting a mouse gene into natural mousepox, the researchers had created a superlethal, vaccine-resistant pox of mice.

If a pox that crashes through a vaccine could be made for mice, then one could probably be made for men.

"My God, Peter, can you believe what these jackasses have done?" Moyer blurted.

Jahrling stared at the poster. He got the point of it right away: the Australians had engineered a poxvirus that could overwhelm the vaccine, and they'd done it by putting a single gene from the mouse into the virus. One mouse gene into the pox. Child's play. "Holy shit," he said.

"This virus just mowed down these immunized animals," Moyer whispered in a low voice to Jahrling, staring at the mouse man from Australia, who was looking rather hopefully at them, like a salesman without any customers. But the two Americans drifted away. "If I were a bioterrorist, Peter, I would rip that paper down and take it home with me." Moyer glanced back at the Australian. "Maybe that paper should come down right now. It makes me wonder if the vaccination strategy for smallpox would work," Moyer said.

JAHRLING WENT BACK to his hotel room and mentally kicked trash cans around. The poster looked to him like a blueprint for the biological equivalent of a nuclear bomb. People were attending the conference from countries that were suspected of secretly developing smallpox as a weapon, and there was no doubt that genetic engineering was something they were perfectly capable of doing. This poster might give them ideas for how to make a smallpox that could be vaccine-proof. He was especially worried about the Vector scientists. Lev Sandakhchiev was walking around, adding his blue Russian cigarette smoke to the haze in the conference center.

It was late in the afternoon, and there was a bus trip planned to the Pont du Gard, a Roman aqueduct that spans a gorge near Nîmes. Jahrling went downstairs and found

Dick Moyer. They got on the bus together and sat down. Then Moyer spotted Ron Jackson sitting by himself near the back of the bus. "See you later," Moyer said, and he hurried down the aisle and claimed the seat next to Jackson.

"That paper of yours is one of the best papers at the meeting," he said, trying to break the ice.

The bus wound through the beautiful terrain of Languedoc, through olive groves and past limestone cliffs. Moyer found Jackson to be a "nice guy, kind of a shy guy, and a good scientist." They had a talk about how, exactly, the engineered pox had wiped out the immune mice. Moyer was very interested in the exact way in which a poxvirus could trigger a storm in the immune system and overwhelm the vaccine. "Ron Jackson and his group knew what they had done," he said later. "Anybody working in this field would have to be absolutely retarded not to see the implications of it with the vaccine for smallpox. They're professionals, and they saw it. They agonized over publishing their experiment. But I still can't believe they published it." A vaccine-resistant smallpox would be everyone's worst nightmare come true. We could be left trying to fight a genetically engineered virus with a vaccine that had been invented in 1796.

THE AUSTRALIAN RESEARCHERS were working for the government, and they had asked officials what they should do. Information travels fast via the Internet. Word could leak out about their experiment, even if they didn't publish it. Putting the IL-4 gene into a poxvirus was such simple work that a grad student or summer intern could probably do it. Virus engineering had become standardized, and there were kits you could order in the mail for doing it. It was getting easier to alter the genes of a virus all the time, and poxviruses were just about the easiest viruses to engineer in the laboratory.

Ron Jackson and his colleagues—principally, a molecular biologist named Ian Ramshaw, who had done the technical work of constructing the virus—talked it over with one of the leading eradicators of smallpox, the Australian pox virologist Frank Fenner. Fenner had done some of the early and important research on mousepox virus, and he is the principal author of the Big Red Book—*Smallpox and Its Eradication*. He advised them to publish. He felt that there were reasons to think that IL-4 smallpox—smallpox with the human IL-4 gene spliced into it—might not work the same way as IL-4 mousepox did in mice. Furthermore, he felt that an engineered smallpox that did spread through vaccinated humans would not be useful as a biological weapon because it would kill too many people too fast, and so would not spread well, in his opinion, and it might kill the people who made it. Fenner also believed that a terror group or a nation would need to test the engineered smallpox on human subjects in order to be sure it worked. That was a difficult hurdle, he reasoned.

As for Jackson and Ramshaw, one impulse for publishing their work seems to have been simply to remind the world that the genetic engineering of virus weapons was something quite possible. They wanted to warn the community of biologists to stop pretending the problem didn't exist, and to start discussing it and dealing with it.

The Jackson-Ramshaw paper was published, with a small burst of publicity and media attention, in the *Journal of Virology* in February 2001. At that point, the technique for engineering a presumably vaccine-resistant super mousepox became available worldwide on the Internet.

The Jackson-Ramshaw experiment provoked an uneasy reaction in the American intelligence community. CIA biologists were apparently aware of the paper, since it pointed to a vulnerability in the government's plans to assemble a stockpile of vaccine. The paper was discussed at the National Security Council. One member of the NSC believed that the Australian scientists had intentionally published their experiment out of scientific pride. This was an unreasonably cynical view of Australian scientists, but it reflected the unease with which the intelligence community viewed the possibilities for genetic engineering of virus weapons.

After giving a couple of interviews with journalists, Jackson and his group decided to let others do the talking for them. Dr. Annabelle Duncan, an Australian government scientist, argued that the researchers had done nothing wrong and that unexpected findings are a normal part of science. "I got especially rabid e-mail from people in the United States," she said. "But it would have been silly and dangerous not to publish the paper, because there would have been an implication that we were doing something harmful." She maintained that the group had been surprised by the result and had never thought the immunized mice would die, and this seems true. In essence, the Jackson-Ramshaw team had had a laboratory accident with an engineered virus and had chosen to tell the world what had happened.

A MONTH LATER, officials at the CDC gave the U.S. Army permission to try a second experiment to see if, somehow, they could create a monkey model of smallpox. Peter Jahrling put Lisa Hensley in charge of the experiment.

A Slight Argument

AT EIGHT O'CLOCK in the evening, Peter Jahrling was in his living room, packing a battered suitcase. The sun had set, but the birds were still singing, and the sky glowed with spring. Jahrling had to catch a flight to Atlanta. The Jahrlings' master bedroom is small, and his wife, Daria, had told her husband that she did not want him packing there. "That suitcase of yours has been God knows where, like Siberia. You drag it down streets where dogs walk," she said. "I don't want that thing on our bed."

So he was packing the suitcase on the rug in front of the television set, which was on, although nobody was watching it. Daria was gathering the children's laundry from their rooms, walking briskly around the house with a plastic laundry basket. Their five-year-old daughter, Kira, was rolling on the couch in a bunny suit, drawing on a piece of paper with a crayon.

Daria paused briefly, holding the laundry basket. "Peter, how long are you going to be gone this time?" She is a casual person with an honest way about her. She teaches English at a local high school: Shakespeare and T. S. Eliot and the Imagist poets.

"It depends on how it goes," he answered. He added a T-shirt and shorts.

"I thought you didn't go into the space-suit lab anymore. Don't you have people who can do the work for you?"

He added a light blue polyester sport coat to the suitcase. "Frankly, I'm the only one who has the passion to make it all come together right."

Daria carried the laundry downstairs and started the washing machine. Peter was immune to everything—he had been vaccinated for anthrax and smallpox—but she and the kids weren't. She had told her sister that she wished they all had some of Peter's blood in them. She went back upstairs.

Kira hopped off the couch and ran over to her father, holding her paper. "Daddy, I need a clipboard."

He went into his office and got a clipboard. She hung a picture on it and showed it to him.

"Hey, that's nice, Kira."

"Go brush your teeth, baby," Daria said to Kira. Kira buzzed off to the bathroom.

"I'm going to miss her."

"You never see her. You usually don't get back from work until she's in bed."

"All I can say is, there are reasons for coming up with countermeasures to smallpox. We all know that crazies exist."

A lot of their communication was nonverbal. She gave him a smile that was a mixture of impatience, annoyance, and wry amusement, a look they knew meant, You live in Peter's world.

He tucked Kira in bed and read her a story and arrived in Atlanta at midnight.

Chaos in Level 4

THE MONKEY-MODEL team stayed at a hotel in the suburbs, not far from the CDC. At sunrise they were drinking coffee and eating bagels, scrambled eggs, and fruit in the hotel's café. The monkey-model team consisted of Peter Jahrling, John Huggins, Lisa Hensley, and an Army veterinary pathologist, Lieutenant Colonel Mark Martinez. There was also an animal caretaker named James Stockman and two veterinary technicians, Joshua Shamblin and Sergeant Rafael Herrera. A separate science team, headed by a biologist named Louise Pitt, ran the Monkey Cabinet. This was big biology—expensive and complex. Everyone in the room was keyed up.

Lisa Hensley wasn't a morning person and never ate breakfast. She bought a Diet Coke and drove with Sergeant Herrera to the CDC in a rented car. It was a cool, pleasant morning, and the sun was flashing through chinkapin trees and loblolly pines, and the air held scents of Georgia summer. They drove down a hollow and up a hill, turned in to the CDC campus, and showed their identification badges to a security guard. The badges were marked "Guest Researcher."

They walked through a security door and crossed an open area, went through another security checkpoint, and arrived inside the Maximum Containment Lab. The MCL is a six-story building but does not seem large; it is embedded in the side of a hill, and three of its stories are partly belowground. It is attached to a larger structure known as Building 15. The MCL has a line of purplish smoked-glass windows that make the building look like an aquarium. There were television cameras and armed guards around. The variola freezer had been removed from its normal hiding place or places, and the security people had a live camera watching the freezer inside the hot zone.

CDC officials had decided that the Army people could work in a corridor of the sub-subbasement. The Army people felt they were getting a bit of a hazing, for it was clear that not everyone at the CDC was happy to have them there, working with smallpox. As an institution, the CDC was proud of the leading role it had played in the Eradication, and there were undercurrents of feeling around the CDC that it was just not right to be warming up variola and doing experiments with it.

The Army's work area consisted of three small desks lined up in the corridor, illuminated by basement windows that looked out on the wheels of parked cars. Hensley sat down at a desk, pulled the Diet Coke from her bag, popped it open, and sipped it.

The others arrived, but there weren't enough desks, so they stood, drinking coffee from foam cups. The animal caretakers were going to go in first, to feed the monkeys. Hensley waited for a while and then went up three flights of stairs and through another security point to an entry door that led inward to the smallpox. The MCL was divided into two separate hot zones, east and west. She pushed through a small door into MCL West and into a small locker room, where she undressed.

There was a circular scar on her upper left arm—the

site of a fresh smallpox immunization. She pulled a green cotton surgical jumpsuit from a shelf and buttoned up the front. The fabric was faded and tore easily: it had been sterilized in an autoclave many times. Another shelf held athletic socks that had been sterilized and were crispy and brownish. She rummaged around for a pair that seemed less crispy. Barefoot and holding the socks, she walked through a wet shower stall and opened a door. It led to a supply closet. She walked through the closet, pushed open a door, and entered the space-suit room.

It was a Level 3 room, close to the hot side, jammed with blue space suits hanging on hooks. Each suit was marked with the name of its owner. Most of the suits belonged to CDC scientists. They had seen hard use—the seats of some of them were patched with black tape. (They tend to develop holes in the buttock area when you sit down.)

Her space suit was brand-new. She really liked that new-space-suit smell. She snapped on surgical gloves, taped the wrists to the sleeves of her scrubs, and carried her suit back into the supply closet, where she sat on a box and put her legs into the suit. She stood up, pulled the faceplate down over her head, and closed the front seal, which snapped shut automatically. She selected an air regulator—a steel canister with a shoulder strap. She slung it over her shoulder and plugged the regulator into her suit.

There was a stainless-steel door on the inward side of the room that had the red biohazard symbol on it. She shuffled into the air-lock decon shower, closed the outer door, opened the inner door, and stepped through to the hot side. She was in a small room where galoshes were sitting on the floor—the boot room. She stepped into a pair that looked about her size. The galoshes were to protect the feet of her suit from developing holes. Then she pushed through a swinging door into the main room of MCL West.

The main room was forty feet long, and it was in the shape of an L. The walls were covered with brilliant white tiles, and the light was bright. Red air hoses dangled in coils from the ceiling. An array of freezers stood along one wall, one of which was the smallpox freezer. Hensley started moving through the room. You didn't exactly walk in Level 4, you shuffled. She pushed through a door into a lab room. This would be her workplace for the duration of the experiment. She stood up on tiptoes, pulled down an air hose, and plugged it into her regulator. There was a roar, her suit pressurized, and dry, cool air washed past her face. She spent the morning setting up test kits, getting ready for the awakening of variola.

On the far end of the main room there was a heavy steel door, and beyond it was the animal room, which was now crowded with people in space suits. The room contained four banks of monkey cages. The monkeys were calm, not vocalizing much, since they had been living in Level 4 for weeks, and they had grown used to being around humans wearing space suits. Each bank of cages had a plastic tent over it, to keep smallpox from spreading from one bank to another, in case any monkeys did develop smallpox. There were eight monkeys in the cages. The monkeys were crab-eating macaques from Southeast Asia. They had grayish-brown fur, pointed ears, and sharp, canine fangs. Jim Stockman, the animal caretaker, had fed them a breakfast of monkey biscuits. They had eaten some of their biscuits and had thrown others around the room. Stockman had cleaned up the mess. All the cages had brass padlocks on them—a crab-eating monkey can figure out any latch.

Mark Martinez, the veterinary pathologist, was in the monkey room, too, getting things set up. Martinez is a soft-spoken man in his forties, with brown eyes and wire-rimmed glasses. Some years earlier, he had attended Airborne school at Fort Benning, Georgia. One day, he had been walking around the base and had found a graveyard for dogs, overgrown with weeds. Among them were brass plates and slabs of stone. Each marker displayed the name of a dog, with its dates of birth and death. They had been killed in action during the Vietnam War, had been shipped home, buried, and forgotten. Martinez thought about how many of the dogs had died in combat, perhaps defending their human companions, and he had the graveyard mowed and tidied up, and he cleaned the grave markers. He felt that the dogs had died for their country.

LISA HENSLEY was standing at a work counter, setting up her equipment. She was looking down at her hands, when she noticed that her right outer glove had developed a crack in the wrist. The glove was rotten.

She had no tolerance for bad gloves. Time to get out.

She picked up a bottle of Lysol, sprayed her glove, and headed for the exit. She took off her galoshes and stepped into the decon shower and pulled a handle to start the cycle.

A spray of water, then Lysol came down over her. After it ran for seven minutes, she turned a handle to shut it off. It wouldn't turn off. It was jammed open, and the shower was still running with Lysol.

"Oh, crap," she said. She returned to the hot side and tapped the vet tech, Josh Shamblin, on the shoulder and pointed to the air lock. "It's running. It won't stop." She had to shout, since both of them were wearing earplugs inside noisy suits; it helped if you could read lips.

He said to her, "Get Jim."

Jim Stockman had worked in MCL West before, and he knew how to fix the decon shower. He clambered into the air lock and started banging around in the spray, trying to fix the mechanism.

Peter Jahrling had arrived in the gray area with John Huggins, and they were peering at Stockman through a window in the air lock. "What the heck are you doing?" Jahrling mouthed.

"Fixing it."

Suddenly, the floor drains in the main room began puking up nasty yellow foam. It was dirty Lysol, overflowing from the waste drains.

Rafael Herrera came running out of his workroom, lumbering in his suit, shouting, "We've got a flood in here!"

Hensley went over to the window of the shower and started pounding on the glass. "Jim! Jim! Look!" Mark Martinez and the others were now careening around the main room, their voices sounding dull and faint as they yelled at one another inside their suits. One of them picked up the receiver of a wall phone and called the CDC's Level 4 janitorial services: "We've got a Lysol flood in here! The plumbing is backing up!"

The monkeys probably thought it was pretty exciting.

The shower stopped. Stockman opened the door of the air lock, and more Lysol poured into the room. They found a Sears shop vacuum and ran it around the hot zone to suck up the flood.

It had been a long day. The team drove back to the hotel, and most of them went straight to bed. Hensley stayed up long enough to phone Rob Tealle. She told him that everything was okay, except for a flood in the lab. He had been working on a project to build some houses. The project was winding down, and he was planning to direct his business toward furniture building. It was a brief conversation.

The Awakening

AT EIGHT O'CLOCK the next morning, John Huggins crossed through the main room of MCL West in a blue suit, went somewhere, and retrieved the Smallpox Key. Huggins is a calm, deliberate, chunky man with a pointed nose, tortoiseshell eyeglasses, and dark, wavy hair with a splash of gray at the temples. He went to an array of freezers of various kinds, all lined up against a wall. There were chest freezers, and there were freezers that looked like kitchen refrigerators, and there were several cylindrical tanks made of stainless steel, sitting on wheels, which were liquid-nitrogen freezers. The freezers all had digital displays showing the temperature and status of the freezer.

The liquid-nitrogen freezers were shiny and new and looked a little like nuclear-reactor vessels. Each contained a shallow pool of liquid nitrogen in the bottom. One of them held the smallpox.

The freezers were chained to the wall with monstrous steel chains. Huggins shuffled up to the smallpox freezer, inserted the Smallpox Key into some sort of lock, and an alarm was disarmed, and the lock opened. There was a red panic button on the wall near the smallpox freezer. When the researchers mushed around in their space suits near the smallpox freezer, they worried that they would accidentally bump the button and draw an armed response.

Huggins slid a terry-cloth mitt over his right space-suit glove, and he lifted up the hinged circular lid of the freezer. He pushed it all the way up and back, and it clanged open. There was a whooshing sound, and a cloud of white nitrogen vapor billowed out of the top of the freezer, poured down its sides, and spread out across the floor, running around his legs. He could barely see the white cardboard boxes in the mist inside the freezer. He had about three minutes to get the vials he wanted before the pool of liquid nitrogen in the bottom of the freezer came to a boil and threw up a huge cloud of fog. The boxes were stacked one on top of another on steel racks.

He couldn't see a thing. He reached into the fog and started feeling around. He counted down a certain number of boxes and slid a box out of a rack. He removed the lid of the box, which was still hidden from sight in the fog. The inside of the box was divided into a grid, and with his fingertips, he counted a certain number of columns across, then down a certain number of rows. He pulled five vials out of the grid. He wedged the vials into a plastic rack, lowered the lid of the freezer with a clang, and locked it.

The vials contained smallpox seeds. Each seed was a lump of frozen amplified smallpox soup the size of a pencil stub. Holding the rack of seeds in one hand, he returned the Smallpox Key to its hiding place. He then carried the seeds into another room, where he placed them in a tank full of water that was kept at 98.6 degrees Fahrenheit, the temperature of blood.

While John Huggins was warming the smallpox, the air-lock door opened, and Peter Jahrling stepped through to the hot side. He was wearing a Delta Protection space suit—not a blue suit but a Day-Glo orange one of French design. The other scientists thought Jahrling looked hip—French couture in Level 4.

Jahrling shuffled over to Huggins: "How's it going?"

"Ready in five minutes."

Jahrling's heart was racing. He thought this was what a NASA launch must feel like. It seemed clear that if they failed, the WHO would not allow any more animal experiments with smallpox. And that could slow down the development of new drugs for smallpox for the foreseeable future.

He left Huggins with the smallpox and went down the hall to see what Hensley was doing in her lab. Jahrling shouted: "Are all your tubes labeled, Lisa? Are the tubes lined up in the right order? You want to be ready to go when John brings out the variola."

"Naw, I figured I'd wait till the last minute." She gave him a little grin.

He didn't think it was funny. "You want to leave as little to chance as possible, Lisa."

"Uh-huh. Yes, sir." *Give me a break, Dr. Jahrling,* she thought.

Huggins plucked the seed vials out of the water. The seeds had melted to a pink, milky liquid, which shimmered with faint opalescence, like mother-of-pearl. It was the same opalescence that appears in the pus of human victims of smallpox. He held the vials in the light and tipped them gently, peering into the variola, checking to see if it had melted completely.

The strain was called the Harper. It was collected in 1951 from an infected American soldier whose name may have been Harper, and it had somehow ended up in the Japanese national collection of smallpox, under the control of Dr. Isao Arita, one of the leading eradicators. The Harper was delivered to the CDC after the Eradication. Researchers believed it was a very hot strain.

John Huggins opened the vials and sucked the Harper from them with a pipette and dribbled the liquid into four syringes. As he loaded the syringes, he got a sweaty-palmed, nervous feeling. He was loading each syringe with a billion particles, maybe three hundred million cases of variola in one syringe, enough to toast North America. It was *hot* material, and it made his heart race. Huggins had handled amplified seed stock of variola before—indeed, he had grown these seeds in virus cultures in the MCL during the previous few days—but no matter how many times he handled liquid seed smallpox, he could never feel calm about it. You could just feel the explosive, infective power in those syringes. He had been vaccinated many times, but the slightest needle stick from a syringe full of amplified Harper would be likely to blow through his vaccinations like a bullet through toilet paper.

After he had loaded four syringes, he loaded an additional four vials with the Harper seed liquid. He put the vials and the syringes on a tray and carried them into the animal room, walking slowly, watching where his feet were going, and holding the tray as if it had an atomic bomb on it.

THE MONKEY ROOM was jammed with Army scientists and technicians wearing blue suits, waiting for the arrival of the Harper. They were hooked up to the air lines that coiled down from the ceiling. The monkeys were vocalizing, and eeks and calls sounded in the room. In the upper left cage of the first bank, a large male crab-eating macaque, Monkey C099, was watching the people. He was an alert animal, calmer and more inquisitive than the others.

Stockman and Martinez had taken note of the monkey. They had many years of experience with animals, and they were aware of differing "characters" among them. Monkey C099 had a pale muzzle with pinkish-white skin that was free of facial hair, which was unusual for a crab eater. It gave him a more human appearance than some of the other monkeys. He was a leader type, more confident of himself, and one of the largest males in the group. He had big, sharp, canine teeth. He was not a monkey to mess with.

Stockman freed some latches on Monkey C099's cage and pulled a squeeze panel forward, which moved him toward the front of the cage. Stockman worked slowly and very gently, trying not to upset the monkey. While Stockman kept the monkey pressed against the front of the cage, Josh Shamblin took a syringe and gave the animal a shot of an anesthetic, Telazol, in the thigh.

They waited a few minutes. Monkey C099 settled down, and he almost went to sleep. Then Shamblin unlocked the brass padlock and lifted the animal out, holding it under the arms. He shuffled across the room, carrying the monkey in front of him.

Meanwhile, Huggins gave some vials of smallpox to Louise Pitt, who was in charge of the Monkey Cabinet. She loaded the liquid into a device that would make a mist inside the aerosol cabinet. Shamblin handed the monkey to Pitt, and she placed him on a table inside the chamber, lying on his back.

Then Pitt gave the thumbs-up, and one of her team started a blower running. Harper smallpox was blown into the air around the monkey's head. The monkey breathed one hundred million particles of Harper. The monkey yawned, exposing his fangs. He was loopy with the anesthetic.

The monkey was also due to get a load of smallpox straight into the bloodstream. Shamblin took an IV needle and inserted it into a vein in the monkey's thigh. He attached a tube to it, and he took up one of the syringes full of Harper, uncapped it, and very carefully fitted the syringe into the tube. He injected the monkey with around one billion infective particles of Harper smallpox. There was a pause. Josh Shamblin glanced around, catching people's eyes to make sure that he had the attention of the room. The roaring of air in people's suits was too loud for speech. He was letting everyone know that he was about to pull the contaminated needle out of the animal.

Everyone stopped what they were doing and froze, and some people stepped backward. When he was sure the room was under control, Shamblin pulled the needle from the monkey's thigh. The steel glinted, slick with monkey blood, and every human eye in the room watched it. It was perhaps the dirtiest needle in the history of variola's entanglement with the human species. Without capping it—he wasn't going to get his fingers anywhere near that bloody point—he took two steps across the room and dropped it in a biohazard container. The biohazard waste containers were cooked in autoclaves inside the Maximum Containment Lab, and only after they had been sterilized were they removed from Level 4.

Then they carried the big, sleepy, pale-faced monkey back to his cage, and they repeated the process three more times with three more monkeys.

The next day, they inoculated four more monkeys with a different strain of smallpox, the Dumbell 7124, which the researchers usually called the India strain. It had been collected in 1964 in Vellore, in southern India, by a British smallpox researcher named Keith Dumbell. Three years later, in Vopal, India, Soviet scientists collected the strain known as India-1, which became their strategic-weapon strain. The Russian government has refused to share the India-1 strain with anyone, but Jahrling and his group believed that their India might be similar to the

Russian India-1. They regarded it as the hottest smallpox that anyone outside Russia could obtain.

This test was different from Jahrling's previous experiment, when his group had failed to infect monkeys with smallpox. They had used a lower dose then, and they had given the virus to the monkeys through the air. This time they also put it into the bloodstreams of the monkeys, and they used a higher dose. Jahrling felt that if a billion particles of smallpox didn't give a monkey a disease of some sort, then nature would be telling us that variola was not going to go into any species except man.

They kept the monkeys under observation, wondering what would happen. The Harper and the India might start multiplying in the monkeys or it might not. If the monkeys became sick, no one knew what the disease would look like. It was impossible to say what variola would do.

THE
DEMON'S EYES

Down

AFTER THE MONKEYS had been inoculated with two strains of smallpox at the CDC, Peter Jahrling and John Huggins flew back to Maryland. They left Lisa Hensley in charge of the experiment, assisted by Mark Martinez. They were supported by Jim Stockman and Josh Shamblin. The team members settled into a routine. They arrived at the CDC at seven o'clock in the morning and checked in through security. Stockman and Shamblin would immediately put on blue suits and go into MCL West. The monkeys threw their bedding out of the cages every night. The bedding consisted of balls of paper, which the monkeys seemingly enjoyed throwing around the lab. Stockman would clean up the paper and clean out the cages and give the monkeys their biscuits, assisted by Shamblin, who also readied things for blood tests of some animals. Down in the basement corridor, Hensley and Martinez would sit at the desks with their laptop computers, catching up on e-mail, drinking coffee and Coke.

It had become increasingly clear to Hensley that Peter Jahrling had pulled a fast one on her. She saw, now, that he had intended to put her in charge of the experiment, but he hadn't told her so. She thought it was funny—Dr. Jahrling had been afraid she would refuse to work with smallpox if he asked her. In fact, she longed to get back to work with Ebola. She felt lonely in Atlanta. She missed Rob Tealle, but when she was in the throes of a big science project, she tended to put him in a different compartment of her life. Going into Level 4 with smallpox seemed to her a little like being an astronaut and going into orbit for months. The world fell away as you entered the air lock, and you focused on the work at hand. You lived with breathing equipment day after day, and you watched your hands every moment.

She was always the last to enter MCL West each morning, and she always checked the animals. Eight of them had been exposed to smallpox, but there was no outward sign of illness in them.

Monkey C099 seemed gentler than the others. The scientists got into the habit of feeding him treats—marshmallows, cotton candy, and popcorn. This was nice for the monkey, and it didn't affect the experiment. Holding a tuft of cotton candy, Stockman would go up to a cage, and a hand would whip out of the cage, almost faster than the eye could see, and the tuft would vanish in the monkey's mouth. Then the hand would reach from the cage again, as the monkey asked for more.

Each day, the scientists would give some of the monkeys a shot of anesthetic and stretch them out on a table in a room next door, to examine them, and Shamblin would extract blood samples. Mark Martinez would fill a series of Vacutainer tubes and hand them to Lisa Hensley, who labeled them and took them into her lab and ran dozens of tests on the blood, looking for any changes.

On Day Two of the experiment, Hensley detected the DNA of smallpox in the blood of the monkeys. It had not been there before. This meant that the virus was almost certainly growing in the monkeys.

Hensley went home each night to her room at the

hotel. If there was time, she jogged in a nearby park, or sat by the pool with other members of the team, who drank beers and unwound, or they would go out and get pizza. She usually didn't drink alcohol during a Level 4 project. Frequently, she warmed a Healthy Choice dinner in her kitchenette, spread her papers and laptop out on the sofa, and worked on her Ebola project data, sometimes until late at night. When she had time, she would call Rob Tealle, or she would have a chat with her parents. She and Tealle had been wondering if they should get married. They had been living together for quite some time, and Hensley felt a desire for a home life pulling on her. Her twenties were passing, and she wanted to have children someday. Her older sister had become a mother and was happy and fulfilled with her child. Hensley kept photographs of her little niece above her desk at USAMRIID.

JUNE 4th, 2001, was Day Four for the four monkeys that had been exposed to the Harper strain. It was Day Three for the monkeys that had been exposed to India. Lisa Hensley and Mark Martinez arrived early in the morning, put their laptops on the desks in the basement, and started swapping the one phone jack to send e-mail. Jim Stockman put on a blue suit and went in to take care of the monkeys. A few minutes before eight o'clock, a telephone in the hallway rang, and Hensley answered it. It was Stockman, calling from MCL West.

He was shouting through his faceplate, "We've got two monkeys dead in here! Another one is going down!"

She thought Stockman was joking. "Oh, yeah, like whatever," she blurted, but Stockman was a very serious man, and her heart went *wham,* and she could feel massive amounts of adrenaline kicking in. *Monkeys down.*

Martinez jumped up and began to move fast. He wanted to collect clinical samples from the dead monkeys, and hurried to put on his space suit. As soon as he had left the locker room, Hensley followed him in.

Martinez went into the monkey room and looked through the plastic tent into the cages. There were two dead monkeys, hunched up, and they had spotty, starlike red spots all over their skin. He thought, *Oh, my God.* The monkeys were speckled—he could see tiny pinpoint hemorrhages all over their faces. The pinpoints were especially dense across the monkeys' eyelids, flanks, and insides of the thighs. This was a flat rash, no pustules. The animals had flat hemorrhagic smallpox. *My God, it was bloody.*

The two dead monkeys were both in the India group. There had never been an animal known to be killed by *any* strain of smallpox. This was the first time anyone had seen variola amplified fatally in any species other than humans.

Feeling somewhat overwhelmed but extremely eager to find out what the India strain had done to the monkeys, Martinez got a pole and gently touched the dead monkeys. He wanted to make certain they were dead. A not-quite-dead monkey that is hot with India smallpox and has canine teeth would be an exceedingly dangerous animal. The touch of the pole revealed that the monkeys were stone dead. One of them, a smaller male designated C171, was in rigor mortis.

Martinez was the team's pathologist. He wanted to do posts on them fast—he wanted to see tissue. He examined the dead animals' eyes. They were normal-looking; there was no sign of blood, as there is in humans with bloody smallpox. He decided to do a necropsy—a postmortem exam—of the heavier male, Monkey C115. He carried the monkey into the necropsy room, laid it on a metal table, and assembled his tools. He closed the door behind him. Animal-use laws prohibit any necropsy or surgical procedure on an animal within the sight of other animals of the same species.

LISA HENSLEY went straight to the necropsy room without stopping to see what was happening with any of the other monkeys. She wanted to get on the necropsy fast.

Martinez had already started by the time she arrived. The monkey was opened on the table; its abdominal cavity was wide, and it was puddled with free hemorrhages—the same thing that happens with human hemorrhagic smallpox. There were hemorrhagic spots all over the internal organs, especially the intestines.

Martinez lined up a row of plastic bins along the edge of the metal table, and he began filling them with samples of the monkey's organs. He worked very fast.

Hensley's heart was pounding. There was an emergency telephone hanging on the wall near the monkey cages. She called Jahrling, reaching him just as he arrived at his office at USAMRIID.

Jahrling started shouting over the phone at her. She could barely hear his voice through her earplugs and the roar of air in her suit. He wanted her to call him from the MCL and report whatever she and Martinez saw, all through the day. He sounded hyper.

The monkey's stomach was bloody, wrecked by the smallpox. The lungs were bloody and speckled by hemorrhage. The liver was necrotic—mostly dead. The virus had gone everywhere inside the monkey.

She was face-to-face with variola major for the first time in her life. Until she had seen this hemorrhagic monkey, she had had no idea how powerful the virus was, how truly frightening. It was scarier than Ebola, much scarier, because it was a virus that was superbly adapted to humans, and it spread in the air. Ebola spread only by direct contact, and it was not well adapted to humans. Here, variola would be coming straight into the air out of the animal's body cavity.

"Lisa!" Martinez shouted.

He handed her a plastic bin containing a half-dollar-sized lump of dark meat.

"What the heck's that?" she asked.

"Spleen."

The spleen was a mottled, cloudy, ultraswollen ball—and it was mostly dead. She picked up two scalpels, one in each hand, and bent over the sample in an awkward stance, holding her body back and away from the countertop, with her elbows out. This piece of spleen would contain several million human deaths' worth of variola. She cut delicate bits, mincing the tissue. *This spleen is a moosh,* she thought.

She worked quickly, because Martinez was in a flurry of cutting, and the samples were piling up fast. She stood at a little counter opposite the necropsy table. Occasionally, she unhooked her air hose and carried samples of blood or tissue into her lab and processed them, spinning the blood in a centrifuge, looking at it under a microscope, doing red-cell counts and white-cell counts. She was hurrying back and forth, her hands full of amplified India blood.

The day dragged on, and the first necropsy took hours, because what they saw in the monkey was new to science. Around noon, they considered taking a break. They wanted to get something to eat and go to the bathroom, but they also wanted to get moving on the second dead monkey, which was still lying in its cage. They decided to just keep going.

Meanwhile, word had been traveling fast around the CDC that monkeys were down in the MCL—down with variola. The emergency telephone on the wall of the monkey room rang steadily. Hensley took the calls. People were calling from all over the CDC, and Jahrling and Huggins kept calling from Fort Detrick. Whatever their reservations might have been about Army people working with smallpox, the CDC people were getting excited, too. A CDC expert in Ebola named Pierre Rollin volunteered to help, and he arrived in MCL West with some compounds that could be used to prepare the tissues for viewing under an electron microscope.

In the cage on the upper left of the bank of cages in the monkey room, the inquisitive male with the light hair and unusual face, Monkey C099, was taking the scene in calmly. He seemed flushed. Maybe he was going to go down, too. Another infected monkey was looking very sick and was sitting down. Most nonhuman primates do not like to sit down in the presence of humans and will get to their feet if a person is near them. But if a monkey is ill, he will sit down in the presence of a person. The sick monkey hugs its knees and watches people, and it won't eat. A monkey never lies down in front of a human if it can possibly help it. If a monkey is very sick, it will lie down when people turn their backs, but if anyone looks at the monkey, it will sit up again.

The sick monkey was huddling and holding its knees, and there were starlike speckles on its eyelids. When Hensley turned her back, the monkey lay down in its cage.

Martinez was standing next to Hensley. He shouted, "It's going to kill all of them fast. I'll bet we'll be out of here in two weeks."

"You wait and see," she shouted back. "I'll bet there will be one survivor. And we're going to be here for a long time, Mark."

They carried the second dead monkey into the necropsy room. Martinez was in good physical condition, but the strain of the necropsies was already starting to wear on him. He was a white-water kayaking instructor, yet doing pathology work in Level 4 with animals dying of hemorrhagic smallpox was pushing the envelope of his sense of physical control of his surroundings. The work was intense. Every move had to be right. You had to watch your hands, and you had to be superaware of who was around you and what they were doing.

Martinez found a chair, carried it into the necropsy room, and sat on it while he performed the second necropsy. He found that it helped him focus. Hensley had to keep running samples back into her lab, so she remained on her feet. Her back began to hurt, and she was freezing cold. It was the cutting posture that strained her back—bent over in that hunch, elbows out, body held back and away from the scalpel blades, while she took tiny slices of hot tissue. Something about the dry air inside the suit and the air-conditioning in the MCL was enough to practically give you hypothermia, even in summer in Atlanta. Her boots were thin rubber, and she could feel the concrete of the floor through her socks.

They finished the second necropsy at three o'clock in the afternoon.

"Let's shower out after this," Hensley said, and Martinez nodded.

But when they returned to the monkey room, they were shocked to find that the third monkey had already died. This monkey was the first to die of the Harper strain.

They forgot about taking a break and did the third necropsy. The work dragged on for hour after hour, and the sun began to set. There were no windows in the MCL to the outdoors, but a line of windows in the main room looked into a glassed atrium in Building 15. The light in the atrium went dark, as people went home from work. Martinez and Hensley had been in their space suits since eight o'clock that morning. They hadn't eaten anything, and they hadn't been able to make a pit stop in the rest room. The air in the suits was bone-dry, and they were dehydrated and thirsty.

At about eight o'clock at night, Martinez suddenly unhooked his air hose and signaled to Hensley that he was going to exit. She thought he was having trouble with his air supply. He ran out of the room, hurrying for the air lock. The trouble was with his bladder.

He stood in the chemical shower in the air lock in agony. The shower cycle was automatic and took nine minutes, and he couldn't get out until it had finished. Then he ran through the gray zone, tearing off his suit, on his way to the bathroom.

The team returned to the hotel that night and sat around the swimming pool, feeling a little stunned. Businesspeople passed by, talking about sales and deals; a man shot baskets on a little court near the pool; children yelled in the water. Life went on. The purpose of the work in the hot lab was to protect these people from variola, people who probably never thought about the disease and had little idea what it was.

Hensley went to her room and lay down flat on the floor and looked at the ceiling, trying to relieve the pain in her back. This was dramatic work that was going to get international attention. It might be published in some big journal like *Science* or *Nature*, and it was likely to upset the smallpox eradicators.

Harper

TWO DAYS AFTER the three monkeys died, Monkey
C099, the handsome monkey, had tiny pimples spreading
across his thighs, although he didn't seem very sick. They
anesthetized him, laid him down on the necropsy table,
and inspected him. They opened his mouth and found
several small pustules on his palate and inside his lips.
They used a swab to take a sample of saliva from the back
of his throat. They wanted to find out if the virus moved
into the air from the back of the monkey's mouth, as it
seems to do in humans. They returned him to his cage,
and he woke up shortly afterward. He seemed perkier
than the very sick ones.

In the next few days, C099 developed classical ordinary smallpox. It looked to Hensley and Martinez exactly
like human smallpox, which meant that it could be a
model of smallpox that the Food and Drug Administration might accept.

As the pustules enlarged and spread over the monkey's
face and hands and feet, the team saw that the pustules
had dimples in them. This was a centrifugal smallpox
rash, just like the ones humans get. Martinez brought an
underwater camera into the lab, and he photographed
the monkey. He had to use a waterproof camera because

in order to take it out of Level 4 he had to submerge it in a dunk tank full of Lysol for half an hour.

The pustules clustered thickly around the animal's extremities, just as they did in people with smallpox. The scientists began to feel sorry for him. They named him Harper, after the strain he had received.

Harper had one hundred and fifty pustules; they counted them while he was unconscious on the table. Hensley found the classic form of the disease more awful to look at than the bloody form, and this pale-faced monkey reminded her of a human child. She didn't doubt that animal research was needed to save human lives—a prime example being research to find drugs that would be effective on HIV. The smallpox experiment had been reviewed and approved by the USAMRIID and CDC animal-use committees. Any animal that was clearly dying had to be sacrificed right away, and painlessly, so that its suffering would stop. But Harper was not dying. He was experiencing a form of agony that was the heritage of humanity, not of monkeys.

On the morning of June 7th, Harper was huddled in the back of his cage, visibly much sicker. The worst of it was his hands. The pustules of variola had erupted there.

The hand is a symbol of humanity, part of what makes us human—the hand that carved the Parthenon, painted the hands of God and Adam on the ceiling of the Sistine Chapel, and wrote *King Lear* was the only hand that had known smallpox. That same hand had now given the disease to a monkey.

The scientists were watching Jim Stockman, too. He was a serious man in his fifties who had worked with animals for his entire career and he was naturally gentle around animals. They felt that he might be having a difficult time watching Harper come down with smallpox. The monkey was getting dehydrated because he could hardly swallow. Stockman went to a drugstore and bought a bottle of grape-flavored Pedialyte—a fluid replacement that is often given to children who have diarrhea—hoping it would appeal to Harper. Hensley and Martinez prowled the breakfast bar at the hotel, picking over the fruit salad, taking red grapes, peaches, slices of mango and soft banana, tucking the fruit into foam coffee cups, and bringing it into MCL West to see if Harper would want any of it.

Stockman poured Pedialyte into a syringe that had a long plastic tube on it. The monkey took the liquid in his mouth. He seemed to trust the people in the space suits. Shamblin and Stockman pulped up bits of fruit and put them on a tongue depressor and offered them to Harper. He couldn't chew, but he mouthed the mush and swallowed it. He had pustules on his haunches, and Mark Martinez got a soft pad and managed to slide it under the monkey, to help him sit more comfortably. They discovered that he liked the red grapes best of all, and Hensley would clean out all the grapes from the breakfast bar. Stockman bought bags of marshmallows, and Harper managed to chew and swallow them.

Harper had gone semiconfluent across the face. He began to reach the stage of early crust, the most dangerous stage of human smallpox, when the cytokine storm goes out of control. Around June 10th, when the monkey had crusted, Stockman offered him a whole red grape. He reached out, took the grape, and put it in his mouth.

Harper began to seem a little better, and he developed a passion for the grapes. If he noticed that someone had a cup of them, he would hold out both blistered hands and then stuff grapes into his cheek pouches until they bulged with grapes, saving them for later.

• • •

LISA HENSLEY had been phoning Peter Jahrling every day, and the team e-mailed pictures of Harper's face to him. In late June, Jahrling brought some of the pictures to a meeting in Washington at the National Academy of Sciences, where he ran into D. A. Henderson. Members of the National Academy and leading experts on biological weapons were chatting and milling around a coffee machine. Jahrling and Henderson's personal relationship had become tense and sour since Jahrling had begun to argue in favor of keeping smallpox.

Jahrling handed Henderson a color photograph of Harper. "Take a look at this, D.A." The pustules were all over the monkey's face, and they had dimples in them.

Henderson nodded and said something like, "Well, that looks just like smallpox." His point seemed to be that Jahrling didn't need to do experiments with smallpox when monkeypox looked so much like the real thing.

"Well, guess what, D.A.? It *is* smallpox."

According to Jahrling, Henderson shoved the photograph into Jahrling's stomach, turned on his heel, and walked away without a word. Henderson says that never happened.

JUNE TURNED INTO JULY, and Atlanta simmered with heat. Hensley was perpetually chilled in her space suit, and she welcomed the muggy weather when she walked out the doors of the MCL. She had no time for any kind of normal life. Go back to the hotel every evening. Heat up a Healthy Choice dinner. Lie down on the floor. Call Rob. She was making herself less available to him and knew she was doing it, but the experiment was in white water.

Harper had scabbed over, and his health had returned.

They continued to feed him delicacies by hand, but they knew that he wouldn't be permitted to live. The protocol of the experiment required the euthanasia of all animals, in order to gather more data on the effects of smallpox. And there was a biosafety rule that an animal infected with a Level 4 pathogen could not be taken out of Level 4 alive. Smallpox could leave the facility with the animal.

When the day came on which Harper had to be sacrificed, in late July, Jim Stockman announced that he had business to attend to in Maryland, and he would be flying home. Then it turned out that Josh Shamblin suddenly needed to fly home, too.

That night, each of the team members went into the monkey room, one by one, and paid visits to Harper. He had healed almost completely and had no scars. They left him heaps of marshmallows, peanuts, bunches of grapes, and a pear, more than he could eat. The next morning, Hensley and Martinez put Harper to sleep. They used an anesthetic that would cause no pain. The monkey had been anesthetized before, and he would not have found anything unusual about it this time.

Martinez placed Harper, unconscious, on the table and watched him go. He had to note the death formally. Hensley turned her face away.

OF THE EIGHT MONKEYS that were given the Harper or India strains, seven died, six of hemorrhagic smallpox, one of classical pustular smallpox. Harper was the only survivor.

The team infected two more sets of monkeys. In round two, they infected six animals, five of which died. One of these monkeys got pustular smallpox and one of the others developed the brilliant red eyes of human black pox victims. In round three, the final round, they lowered the dose and infected nine monkeys, and none of them got sick at all.

Peter Jahrling felt that the experiments were successful. "We were able to put to rest the myth that smallpox infects no species but man," he said. "We were able to create a disease in the monkeys that approximates the course of the human disease. This means it will be useful for validating antiviral drugs and vaccines for the FDA." He said that the next step would be to challenge monkeys with smallpox and then try to cure them with the antiviral drug cidofovir.

I asked Jahrling about how he justified the suffering of the monkeys in the experiment. "My blood pressure would come down twenty points if we didn't have to work with variola in monkeys," he said. "It really bothers me. The thing is, you look into their eyes and you see they're intelligent. You go into a monkey room at night and you hear them vocalizing, and it sounds like people talking. It really gets to me. But a critical countermeasure to smallpox is going to be antiviral drugs, and the FDA requires testing the drugs on the authentic smallpox virus in an animal. Frankly, I myself could accept and live with an antiviral drug that we've tested in human tissues in vitro"—in test tubes—"and in, say, genetically engineered mice that have been given a humanlike immune system. But testing smallpox on a mouse that has a human immune system isn't going to be acceptable to the FDA anytime soon. Tens of monkeys are going to be sacrificed to this cause, but that is not the same thing as tens of millions of humans with smallpox, and I do believe that smallpox is a clear and present danger. But the truth is that I've been at the point where I really thought I couldn't do this anymore."

Lisa Hensley had experienced grief and sadness over Harper's death, in particular, but she regarded her feelings as a necessary consequence of her job as a public health researcher. "Each of us who does animal research has to weigh in our own conscience what we do," she said to me. "Around twenty percent of the population can't be vaccinated. They're immune compromised, or they have eczema, or they're pregnant women, or they're very young children. That's a large number of people who will have no protection if smallpox comes back. To me, it is not an acceptable loss."

WTC

SEPTEMBER 11, 2001

BY THE BEGINNING OF SEPTEMBER, Hensley had been working with smallpox in a space suit five to seven days a week, without a break, since the end of May. Her parents invited her and Rob Tealle to come with them on a vacation to the Outer Banks of North Carolina, and they accepted. She left Martinez to continue with the smallpox work, which was beginning to wind down.

On the eleventh of September, at 9:00 A.M., Stockman was feeding and checking the monkeys. A CDC smallpox scientist named Inger Damon was taking care of some equipment in one of the rooms. Sergeant Rafael Herrera was working in his suit, listening to music on a radio headset.

Mark Martinez was doing a necropsy of a monkey, and he noticed that Herrera had come into the room. Herrera's eyes were wide, and he mouthed something at Martinez, but Martinez didn't hear it, so Herrera got a piece of paper and wrote: "A plane crashed into the World Trade Center."

"Yeah?" Martinez shouted.

Herrera went out of the room, and Martinez resumed his work. A short while later, Herrera came back, and he wrote on the paper, "Another plane crashed into WTC."

Martinez had to keep working; he was in the middle of the necropsy.

Herrera was listening to developments on his radio headset. He wrote: "Pentagon," "Plane down in PA."

A window in the necropsy room looks out into a hallway. A woman appeared in the window, waving her arms and banging on the glass, and she held up a sign: YOU NEED TO EVACUATE.

A warning had come from high levels in Washington to the director of the CDC, Jeffrey Koplan, that the facility might be a target of a terrorist attack at any moment. It wasn't known in those early hours of September 11th who had carried out the attacks or what other attacks might come. Koplan had ordered an evacuation of all the buildings at the CDC.

Everyone at the CDC knew that the MCL was hot with variola. If it was broken open by the impact of an aircraft or the explosion of a bomb, the smallpox could conceivably escape.

As a lieutenant colonel, Mark Martinez was the ranking officer in charge. He unhooked his air hose and, thrashing in his space suit, ran through the suite, getting everyone's attention, telling them to evacuate. The smallpox freezer was locked and chained, but there wasn't time to do anything about the dead monkey lying on the table.

Martinez ordered people to go into the decon air lock in groups of three. The shower has only two air hoses, so they shared the air. The decon shower filled with mist from the heat of their bodies.

Then a woman appeared in the Level 3 gray area and held a sign up to the air lock door: EMERGENCY PROCEDURES. It meant they had to crash their way out of the Maximum Containment Lab immediately. They wondered if a plane was heading for the building.

They stopped the shower and pulled the DELUGE handle. Many gallons of Lysol splashed over them, and they crashed out of the air lock and got out of the building.

THE
ANTHRAX
SKULLS

Henderson

FIVE DAYS AFTER the fall of the World Trade Center towers, on Sunday, September 16th, at four-thirty in the afternoon, D. A. Henderson was sitting in the den of his house in an easy chair by the Japanese garden, getting no peace from the view.

The telephone rang. It was Tommy Thompson, the Secretary of Health and Human Services, calling from HHS headquarters, on the south side of the Mall. "Can you come to a meeting in Washington?"

"When?"

"Tonight. Seven P.M. We're asking, What's next?" Thompson said. "We'd like you to be there."

Henderson told Nana where he was off to, and he got in his silver Volvo and drove to Washington. It was the end of his plans for retirement. He went to work in Thompson's office and eventually was appointed the director of the Office of Public Health Emergency Preparedness & Response. He became, effectively, the bioterrorism czar in the government, with managerial control over an annual budget that grew to more than three billion dollars. He started getting up at five, taking an early train to Washington, and getting home late at night. He was seventy-three years old. He believed that it was only a matter of time before the bioterror attack that he had long expected finally occurred.

Henderson went to work for the federal government on a Sunday night. The next day or the day afterward—Monday or Tuesday, September 17th or 18th—someone visited a post office or mailbox somewhere around Trenton, New Jersey, and mailed letters full of dry, crumbly, granular anthrax to New York City: to the NBC anchor Tom Brokaw, to CBS, to ABC, and to the *New York Post*.

Into the Submarine

OCTOBER 16, 2001

PETER JAHRLING HAD BEEN in near-daily contact with Lisa Hensley and the monkey team in Atlanta after September 11th, but by the middle of October, he became almost overwhelmed by the investigation of the anthrax attacks, the first large-scale bioterrorism event in the United States.

On the morning of the 16th, the day after it was delivered to USAMRIID, the powder in the letter mailed to Senator Daschle was being studied by John Ezzell, the civilian microbiologist who accepted it from the agents of the FBI's Hazardous Materials Response Unit. But Jahrling wanted Tom Geisbert to get the sample under an electron microscope, and that didn't seem to be happening fast enough. Jahrling met Ezzell in a hallway and said, in a loud voice, "Goddamn it, John, we need to know if the powder is laced with smallpox."

Top Institute scientists were yelling in the halls about an unknown terrorist bioweapon, and the staff rallied. A technician hurried into Ezzell's laboratory rooms and brought out two small test tubes of samples from the Daschle letter. One tube held a milky white liquid. This was from the field test done by the HMRU. The other tube contained a tiny heap of dry particles and a corner of paper cut off the Daschle envelope—the corner was about this size: L. The tubes were inside double plastic bags that were filled with disinfecting chemicals. The technician gave them to Geisbert, who took them into a Level 4 suite called the Submarine.

The Submarine is the hot morgue at USAMRIID. The main door of the Submarine is a massive plate made of steel, with a lever. It looks like a pressure door on a submarine. Pathologists wearing space suits have on one or two occasions used the Submarine for the dissection of the body of a person who was thought to have died of a hot agent, although the opportunity to do this kind of post-mortem exam rarely arises.

Geisbert suited up and went through the air lock into the Submarine, carrying the tubes of Daschle anthrax. He walked past the autopsy room to a small lab. He opened the tube of milky anthrax liquid and poured a droplet onto a slip of wax. Using tweezers, he placed a tiny copper grid on top of the droplet, and he waited a few minutes while the anthrax liquid dried to a crust on the grid. Then he put the grid in a test tube of chemicals, so that any live anthrax spores would be killed. He showered out of the suite, got dressed in civilian clothes, and brought the sample up to one of the scope rooms on the second floor, where he put the tiny grid into a holder and shoved it into one of the electron microscopes, a transmission scope, which is eight feet tall. The scope cost a quarter-million dollars. Geisbert sat down at the eyepieces and focused.

The view was wall-to-wall anthrax spores. The spores were ovoids, rather like footballs but with more softly rounded ends. The material seemed to be absolutely pure spores.

ANTHRAX is a parasite that has a natural life cycle in hoofed animals. An anthrax spore is a seed, a tiny, hard capsule that can sit dormant in dirt for years, until eventually it may be eaten by a sheep or a cow. When it comes into contact with lymph or blood, it cracks open and germinates, and turns into a rod-shaped cell. The rod becomes two rods, then four rods, then eight rods, and on to astronomical numbers, until the fluids in the host are saturated with anthrax cells. An anthrax cell (unlike a virus) is alive. It hums with energy, and it draws in nutrients from its environment. Using its own machinery, it makes copies of itself. A virus, on the other hand, uses the machinery and energy of its host cell to make copies of itself—it cannot live an independent existence outside the cells of its host.

The anthrax cells produce poisons that cause a breathing arrest in their host. Anthrax "wants" its host to drop dead. Anthrax-infected animals can go from apparent health to death with the celerity of a lightning strike. Some years ago, researchers in Zimbabwe found a dead hippopotamus standing upright on all four feet, killed by anthrax while it was walking. The hippo looked as if it had not even noticed it was dead.

The carcass of the host rots and splits open, the anthrax cells sporulate, and a dark, putrid stain of fluids mixed with spores drains into the soil, where the spores dry out. Time passes, and one day a spore is eaten by a grazing animal, and the cycle begins anew.

• • •

GEISBERT TURNED a knob and zoomed in. An anthrax spore is five times larger than a smallpox particle. He was looking for bricks of pox, so he was looking for little objects, searching spore by spore. The task of finding a few particles of smallpox mixed into a million anthrax spores was like walking over a mile of stony gravel looking for a few diamonds in the rough. He saw no bricks of pox. But he noticed some sort of goop clinging to the spores. It made the spores look like fried eggs—the spores were the yolks, and the goop was the white. It was a kind of splatty stuff.

Geisbert twisted a knob and turned up the power of the beam to get a more crisp image. As he did, he saw the goop begin to spread out of the spores. Those spores were sweating something.

The scope had a Polaroid camera, and Geisbert began snapping pictures. He suddenly realized his boss was leaning over his shoulder. "Pete, there's something weird going on with these spores." He stood up.

Jahrling sat down and looked.

"Watch," Geisbert said. He turned the power knob, and there was a hum.

The spores began to ooze.

"Whoa," Jahrling muttered, hunched over the eyepieces. Something was boiling off the spores. "This is clearly bad stuff," he said. This was not your mother's anthrax. The spores had something in them, an additive, perhaps. Could this material have come from a national bioweapons program? From Iraq? Did al-Qaeda have anthrax capability that was this good?

Jahrling got up from the microscope. "I'm going to bring this to the chain of command."

Carrying the Polaroids in the pocket of his gray suit, Jahrling walked across the parade ground of Fort Detrick to the offices of the Army's Medical Research and Materiel Command, which has authority over USAMRIID. It was then headed by Major General John S. Parker, a chunky man with a calm, jovial disposition, wire-rimmed glasses, and a shock of silver hair. General Parker is a heart surgeon. Jahrling walked into his office without knocking. "You need to see this," he said, placing the pictures on the general's desk.

General Parker listened and then asked a few questions. "I want to look at it myself," he said. Jahrling and the general hurried back across the parade ground. It was four o'clock in the afternoon, near the end of a hot, dry October day, and the East Coast of the United States was locked in a drought. Catoctin Mountain looked dreamy and peaceful in the autumn haze. The sun was going down, and the flag in the middle of the parade ground cast a shadow toward the east over heat-scorched grass.

Emergency Operations

LATE AFTERNOON, OCTOBER 16, 2001

GENERAL PARKER AND PETER JAHRLING went by the office of the USAMRIID commander, Colonel Ed Eitzen, and then the three men went upstairs to the scope room, where Tom Geisbert was staring at the anthrax. Geisbert stood up nervously when the general entered and started to explain what he was doing.

"It's okay, I used to run an electron microscopy lab," Parker said.

Parker sat down at the scope and looked. Pure spores. That was all he needed to see. He went out into the hallway and started issuing instructions to Eitzen and Jahrling in a rapid-fire way: We're going to put USAMRIID into emergency operations. We're going to run this facility around the clock. He emphasized that the FBI would be using USAMRIID as the reference lab for forensic evidence from the bioterror event. FBI people would be working side by side in the labs with John Ezzell and other Army scientists. He was going to bring in microbiologists from other parts of his command to help with the work. Parker knew that Washington would be needing as much clear information as possible.

THAT MORNING, a postal worker named Leroy Richmond, who worked at the Brentwood mail-sorting facility in Northeast Washington, D.C., had called in sick. Richmond had a headache, a fever, and pain in his lower chest. He went to bed.

Later in the day, the Postmaster General of the United States John E. "Jack" Potter told his aides to ask CDC officials what should be done about postal workers "upstream" who might have handled the Daschle letter. Officials at the CDC answered that they felt there was no danger to any postal workers. They had a reason for believing that. When they had learned that Robert Stevens and Ernesto Blanco had been exposed through the mail at the American Media offices in Boca Raton, the CDC investigators had taken swab samples in post offices around the area, and they had swabbed the noses of Florida postal workers. They had discovered anthrax spores in the Florida post offices, but no postal workers had become infected. There was no reason to think that postal workers in Washington were in danger.

TOM GEISBERT couldn't keep his eyes off the weapon. He stared at it through the eyepieces of the electron microscope until he noticed that it was eight o'clock at night. He hadn't eaten or drunk a thing all day. He felt like having some breakfast, so he drove out for the double chocolate doughnut with a large coffee that he had been thinking of getting when he had arrived at work. He brought it back to the Institute and continued to work until midnight. He and his wife, Joan, live in Shepherdstown, a long drive to the west. By the time he got home, it was one o'clock in the morning, and Joan was asleep.

THAT NIGHT, a postal worker at the Brentwood mail-sorting facility named Joseph P. Curseen, Jr., began to develop what he thought was the flu while working the night shift, near machines that sort mail. He had a pain in his lower chest and a headache, so he decided to go home. That same evening, one of Curseen's coworkers, Thomas L. Morris, Jr., went bowling. He started to feel sick, and he went home and went to bed to get some rest.

OCTOBER 17, 2001

TOM GEISBERT couldn't sleep. He tossed and turned and looked at the clock: it was four in the morning. He couldn't free his mind of the view in the scope—endless fields of anthrax spores with an unknown substance dripping from them. He got up, took a shower, and left for work. He stopped to buy another double chocolate doughnut and a large coffee, then went to his lab to try to get more images of the anthrax.

AT TEN-THIRTY that morning, the House of Representatives was closed down after CDC people found anthrax spores in mail bins there. About two hundred Capitol Hill workers were told to start taking the antibiotic ciprofloxacin—Cipro. Major General John Parker went to the U.S. Senate, where he met with a caucus of the Senate leadership and their staff. He told them that he'd looked at the anthrax himself in the microscope and that it was essentially pure spores. He would later say, "The letter was a missile. The address was the coordinates of the missile, and the post office did a good job of making sure it got to ground zero."

A HALF MILE away from the Senate, at the Health and Human Services headquarters, D. A. Henderson had been working with Tommy Thompson's staff to get a stockpile of smallpox vaccine created on a crash basis.

There had been fast-paced meetings at the HHS on the subject of this stockpile. Henderson felt that the United States needed one ASAP. Thompson agreed and had just submitted a request to Congress for enough money to create three hundred million doses of smallpox vaccine—one dose for every citizen. The government hired a British-American vaccine company called Acambis PLC to make most of the doses. Acambis's main manufacturing plant is in Canton, Massachusetts. Soldiers surrounded the plant and were stationed inside the American offices of Acambis, in Cambridge. It was thought that a terror attack on the United States with smallpox might be accompanied by an attack on the country's vaccine facilities or an attempt to assassinate Acambis personnel who knew how to make the vaccine. The move to surround the vaccine facility in Massachusetts with military force was done rapidly, in secret, and under apparently classified conditions.

Meanwhile, Daria Baldovin-Jahrling (she uses her maiden name with her husband's) had been getting telephone calls and visits from neighbors. The neighbors knew that Peter was a top government scientist involved with defenses against smallpox, and more than one of them quietly offered Daria money if she could get them some smallpox vaccine. "I don't even know if I can get any for ourselves," she answered them. "If I do, I can't take money for it, and I have to give it to my family first." She was very frightened. "If smallpox was going around Frederick," she said to Peter, "could you get any of the vaccine for the children?"

He told her that if there was a smallpox emergency, their children would get a jab of something in their arms; it might not be the licensed stuff, but it would work. He would make the vaccine himself in his lab if he had to. Yet he couldn't get his mind off the experiment by the Australians, when they had made a vaccine-resistant superpox of mice. What if the vaccine didn't work? He felt the pressure ratcheting up.

WHILE GENERAL PARKER was telling the Senate that the anthrax was pure and the HHS people were asking for money for a smallpox-vaccine stockpile, the FBI decided, sensibly, to get a second opinion on the Daschle anthrax. The HMRU dispatched a Huey to Fort Detrick. Not a few of the FBI's Hueys have bullet holes in them. The holes, which are covered with patches, are left over from combat in the Vietnam War. The FBI had gotten its Hueys used and cheap from the military.

The Huey touched down on a helipad across the street from USAMRIID. An agent went into the building and collected a cylindrical biohazard container called a hatbox. Inside the hatbox, inside multiple containers, was a small test tube of live, unsterilized Daschle anthrax.

The helicopter took off with the sample and thupped westward over Maryland. It touched down in West Jefferson, Ohio, near Columbus, at the Hazardous Materials Research Center of the Batelle Memorial Institute, a nonprofit scientific research and consulting organization. Batelle scientists took the hatbox into a lab. They heated the anthrax powder in an autoclave to sterilize it, and they began looking at it under microscopes.

The spores were stuck together in lumps. They did not appear to be very dangerous in the air—the lumps were too large to float easily or go deep into human lungs. The Batelle analysts conveyed their findings to the head of the FBI Laboratory, Allyson Simons. Their tests showed that the anthrax was not nearly as refined or powerful as the Army people believed.

OCTOBER 18

AT TEN o'clock on Thursday morning, three days after the Daschle letter was opened, Lisa Gordon-Hagerty of the National Security Council conducted an interagency conference call. Such calls were made every morning in the first weeks of the anthrax crisis, and were intended to keep federal officials up to speed. Gordon-Hagerty had her hands full. There were about thirty people listening or speaking on the calls, a cloud of voices. That morning she went around to the various agencies: "FBI, what do you have to report?"

FBI executives in the Strategic Information Operations Center—the SIOC command room—spoke for the FBI. They included Allyson Simons and the head of the Weapons of Mass Destruction Unit, James F. Jarboe. They reported that they were gathering evidence and intelligence on the attacks, and were working closely with the Army to gain a better understanding of the material in the letter that had arrived at the Senate building.

"Army, what are you reporting?" Gordon-Hagerty said.

Jahrling, who was sitting in the commander's office at USAMRIID with Colonel Ed Eitzen, spoke. Choosing his words carefully, because practically the entire executive branch of the federal government was listening to him, he said that USAMRIID had found that the anthrax powder in the letter mailed to Senator Daschle was "professionally done" and "energetic." By "energetic" he meant that the particles had a tendency to fly up into the air if they were disturbed. A key element in the design of a military bioweapon is the weapon's intrinsic energy—the capacity of the particles to fly into the air and form an invisible and essentially undetectable cloud, which can travel long distances and fill a building like a gas.

There were several CDC officials on the call. They were sitting around a conference table in the office of the agency's number two person, Dr. James M. Hughes. Jahrling's voice came out of the box on the table in a tinny way, and it's not at all clear that they understood what he meant by the "energy" of a biopowder. They had not experienced the sight of the anthrax particles floating straight into the air off a spatula—the sight that had prompted John Ezzell to exclaim, "Oh, my God." Furthermore, they did not know much, if anything, about how weapons-grade anthrax is made. Those methods were classified. Perhaps no one had briefed CDC officials on the methods for weaponizing anthrax spores. The CDC officials were public health doctors, and up until then, they had had no reason to learn the secrets of making a biological weapon. To the CDC officials, Jahrling's remarks may have sounded like technical jargon, which it was.

A team of epidemiologists from the CDC was in Washington, working frantically to test five thousand workers on Capitol Hill for exposure to anthrax. They were swabbing the insides of people's noses, concentrating on the people who had been in the Hart Senate Office Building when the Daschle letter was opened. Several buildings on Capitol Hill had been closed down for testing for anthrax spores. The CDC was stretched paper-thin. Many people had essentially stopped sleeping several days earlier, and they were making decisions in a fog of enormous political pressure and exhaustion. The CDC officials did not think that what Peter Jahrling called the "energetic" or "professional" nature of the anthrax suggested that postal workers in the facilities where the letters had been processed might be in danger.

"The significance of the words *energetic* and *professional* were lost on the CDC people," Jahrling said to me. "In my view, at the CDC you have a culture of public health professionals who think of biological warfare as such a perversion of science that they find it simply unimaginable."

The CDC officials on the call asked Jahrling if he could characterize the particle size. This was an important question, because if the anthrax particles were very small, they could get into people's lungs, and the powder would be much more deadly.

Peter Jahrling replied that USAMRIID's data indicated

that the Daschle anthrax was ten times more concentrated and potent than any form of anthrax that had been made by the old American biowarfare program at Fort Detrick in the nineteen sixties. He said that the anthrax consisted of almost pure spores, and that it was "highly aerogenic."

Jahrling now says that he was trying to get the attention of the CDC people, trying to warn them that more people could have been exposed than they realized, but it was like waving to someone across a crowded room. "The CDC people were not reacting much," he said. "I was exasperated. I wasn't getting any response from them when I said the anthrax was highly aerogenic. I was thinking, 'When is this thing going to blow up and get everybody's attention?' "

Jeffrey Koplan, the director of the CDC, was listening on the call but didn't speak much. Months later, Koplan said to me, "If we had known that the anthrax would behave like a gas when it got into the air and that it would leak through the pores of the letters, it might have been useful. But would we have done things differently? You can't say what you would have done differently in the heat and turmoil of an investigation, if only you had known."

The spores of anthrax went straight through the paper of the Daschle envelope and other anthrax envelopes full of ultrafine powder that were mailed, though they had been sealed tightly with tape. It seemed that the anthrax terrorist or terrorists had not planned on having the letters kill postal workers. "They weren't part of the target," as Koplan put it.

Paper has microscopic holes in it that are up to fifty times larger than an anthrax spore. If a pore in the envelope paper was a window in a house, then an anthrax spore would be a tangerine sitting on the sill. If you take a sheet of paper (a page of this book, for example) and seal it against your mouth and then blow against the paper, you will feel the warmth of your breath coming through the paper. This suggests what the anthrax spores did when the envelopes were squeezed through the mail-sorting machines.

AT SEVEN O'CLOCK that evening at the Brentwood mail-sorting facility, technicians wearing protective suits and breathing masks began to walk around the machines, testing them with swabs for anthrax spores. The Brentwood facility was up and running, and there were postal workers all around, working at their places by the machines. One of the workers asked the testers, "How come you aren't testing the people?"

Skulls and Bones

OCTOBER 19, 2001

THE UNITED STATES had been conducting air strikes in Afghanistan for nearly two weeks, and American special forces were operating inside the country. President George W. Bush and his advisers had indicated that the United States considered Iraq to be a sponsor of terrorism, and that Saddam Hussein led "a hostile regime" that the United States would likely target for destruction when it was finished with the Taliban. In the White House, there was extraordinary concern that the anthrax attacks might have been a clandestine operation sponsored by al-Qaeda or Iraq.

• • •

BEFORE DAWN on Friday morning, four days after the Daschle letter was opened, Peter Jahrling put on a space suit and went into the Submarine and got a tiny sample of live, dry Daschle anthrax. He brought it out inside double tubes, for safety, and put the tubes in a radiation pile—a cobalt irradiator—which fried the DNA in the spores, rendering them sterile. He gave the sample to Tom Geisbert, so that Geisbert could look at the dry anthrax in a scanning electron microscope.

Geisbert carried the tube of dry anthrax into his microscope lab, set the tube in a tray, and turned his attention elsewhere. A minute later, he happened to glance at the tube. The anthrax was gone.

Yet the cap of the tube was closed.

"What the heck?" he said out loud.

He picked up the tube and stared at it. Empty. He tapped the cap with his finger, and the particles appeared and fell down to the bottom of the tube—they had gotten stuck underneath the cap, somehow.

He went back to work. A minute later, he glanced over at the tube. The anthrax was gone again. He tapped the cap, and the anthrax fell to the bottom. He stared at the bone-colored particles. Now he saw them climbing the walls of the tube, dancing along the plastic, heading upward.

His assistant, Denise Braun, was working nearby. "Denise, you'll never believe this."

The anthrax was like jumping beans; it seemed to have a life of its own.

He began preparing a sample for the scope. He opened the tube and tapped a little bit of the anthrax onto a piece of sticky black tape that would hold the powder in place. But the anthrax *bounced off the tape*. The particles wouldn't stick. Eighty percent of the Daschle particles flittered away in air currents up into the hood. That was when he understood that the Hart Building was utterly contaminated.

He somehow managed to get some of the particles to stick to the tape. He hurried the sample into the scope room, put it under a scanning scope, and zoomed in. What he saw shocked him.

The spores were stuck together into chunks that looked like moon rocks. They reminded him of grinning jack-o'-lanterns, skeletons, hip sockets, and Halloween goblin faces. The anthrax particles had an eroded, pitted look, like meteorites fallen to earth. Most chunks were very tiny, sometimes just one or two spores, but there were also boulders. One boulder looked to him like a human skull, with eye sockets and a jaw hanging open and screaming. It was an anthrax skull.

The skulls were falling apart. He could see them crumbling into tiny clumps and individual spores, smaller and smaller as he watched. This was anthrax designed to fall apart in the air, to self-crumble, maybe when it encountered humidity or other conditions. He had a national-security clearance, and he knew something about anthrax, but he could not imagine how this weapon had been made. It looked extremely sinister. He started feeling shaky.

He called Jahrling. "Pete, I'm in the scope room. Can you come up here, like right now?"

Jahrling ran upstairs, closed the door, and stared at the skull anthrax for a long time. He didn't say much. Geisbert's security clearance was rated secret, and the details of how this material could have been made might be more highly classified.

Not long afterward, Jahrling apparently went to the Secure Room and had the classified safe opened. He studied a document or documents with red-slashed borders that would appear to contain exact technical formulas for various kinds of weapons-grade anthrax. In the papers, there were almost certainly secrets for making skull anthrax of the type he had just seen in the scope.

Jahrling refers to the secret of skull anthrax as the Anthrax Trick, although he won't discuss it. Could this stuff have been made in Iraq? Could this be an *American* trick? Who knew the Anthrax Trick?

TOM GEISBERT arrived home in Shepherdstown very late. He had been going on maybe three hours of sleep a night for days, but now he had insomnia. He was afraid that his findings about the skull quality of the anthrax meant that it had come from a military biowarfare lab. Finally, he woke up Joan. "I could start a war with Iraq," he said to her. He seemed on the edge of tears. Joan reminded him that he was a scientist and that all he could do was find the truth and report it, wherever it led. "We just have to let the data play out however it plays out," she said. "Other people are working on the anthrax, too."

He did not sleep that night.

LATE ON SUNDAY, October 21st to 22nd, Joseph P. Curseen, Jr., the Brentwood postal worker who thought he had the flu, felt really bad. He had not been to work since Tuesday night. He went to the emergency room at Southern Maryland Hospital Center, where doctors looked at him and sent him home. He was dying, but they didn't see it. That same day, another Brentwood worker, Leroy Richmond, who had called in sick to work earlier in the week, was admitted to the Inova Fairfax Hospital with a presumptive diagnosis of inhalation anthrax, which had been made by an alert emergency room doctor named Thom Mayer. Richmond would eventually survive under the care of doctors at the Fairfax Hospital. That night, at about 11:00 P.M., Brentwood worker Thomas L. Morris, Jr., who had first begun to feel sick during a bowling league event some days before, called 911. He was feeling as if he was about to die, and he told the dispatcher he thought he had anthrax. An ambulance took him to the Greater Southeast Community Hospital, where before nine o'clock the next morning he was pronounced dead. Shortly after Morris died, the Brentwood mail-sorting facility was closed down by order of the postmaster general, and two thousand postal workers were told to start taking antibiotics. Joseph Curseen returned to the emergency room at Southern Maryland Hospital Center on Monday morning and died in the hospital in the early afternoon.

At the mail-sorting facility in Hamilton, New Jersey, a suburb of Trenton, postal workers had been exposed to anthrax, too, because the letters had all been mailed somewhere near Trenton. The Daschle letter had gone through the Hamilton facility en route to Brentwood. A tiny quantity of spores had ended up in the air at the Hamilton mail-sorting facility, and now three postal workers had become infected, as well, two with skin anthrax and one with the inhalation kind.

MEANWHILE IN WASHINGTON, the FBI Laboratory was trying to evaluate the anthrax. On the same day that the two Brentwood workers died, a meeting was held at FBI headquarters involving the Laboratory, scientists from the Batelle Memorial Institute, and scientists from the Army. Batelle and the Army people were doing what scientists do best: disagreeing totally with one another. The Army scientists were telling the FBI that the powder was extremely refined and dangerous, while a Batelle scientist named

Michael Kuhlman was allegedly saying that the anthrax was ten to fifty times less potent than the Army was claiming. Allyson Simons, the head of the Laboratory, was having trouble sorting through the disagreement, and she was apparently not telling the CDC leadership much about the powder, while waiting for more data to come in. One Army official is said to have blown up at Simons and Kuhlman at the meeting, saying to the Batelle man, "Goddamn it, you stuck your anthrax in an autoclave, and you turned it into hockey pucks." He told Simons that she should "call the CDC and at least tell them there is a disagreement over this anthrax." She apparently did not.

The Department of Health and Human Services was not getting briefed about the anthrax to its satisfaction by the FBI. An HHS official who was close to the situation but who did not want her name used had this to say about the Batelle analysis of the Daschle anthrax: "It was one of the most screwed-up situations I've ever heard of. The people at Batelle took the anthrax and heated it in an autoclave, and this caused the material to clump up, and then they told the FBI it looked like puppy chow. It was like a used-car dealer offering a car for sale that's been in an accident and is covered with dents, and the dealer is trying to claim this is the way the car looked when it was new."

THE FBI BEGAN delivering about two hundred forensic samples a day to USAMRIID, frequently in Hueys. Choppers were coming in day and night on a pad near the building. HMRU agents and other FBI Laboratory people began to work inside suite AA3, which ended up being dedicated entirely to forensic analysis and processing samples. The work was done by USAMRIID's Diagnostic Systems Division, headed by an Army microbiologist, Lieutenant

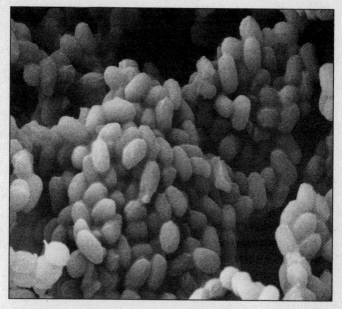

A reference sample of pure anthrax spores, similar in character to the weapons-grade "skull anthrax" in the Daschle letter. The spores are about one micron (one millionth of a meter) across; roughly two hundred spores lined up in a row would span the thickness of a human hair.
(Courtesy of Tom Geisbert, U.S. Army Medical Research Institute of Infectious Diseases.)

Colonel Erik Henchal. The samples were largely environmental swabs—from the Brentwood postal facility, from Capitol Hill, from postal facilities in New Jersey, and from New York City. Each sample was a piece of federal criminal evidence and had to be documented with green chain-of-custody forms. Institute scientists ran ten separate tests on each sample, and every sample ended up matched to an evidence-tracking folder with more than one hundred sheets of paper in it. The hallways of the Institute were jammed with filing boxes full of these folders. In the end, USAMRIID scientists would analyze more than thirty thousand samples related to the anthrax terrorism—far more than any other lab, including the CDC.

One of the many samples was a little bit of anthrax from the letter that had arrived at the *New York Post*. The *Post* anthrax was almost pure spores, like the Daschle powder, but the spores had somehow gotten glued together into glassy chunks. It looked like a glued-together version of the Daschle anthrax.

White House

OCTOBER 24, 2001

EARLY IN THE MORNING, nine days after the Daschle letter was opened, Major General John Parker got a call from Tommy Thompson at Health and Human Services. Thompson had been hearing rumors that the Daschle anthrax was really bad stuff, but he still hadn't heard much about it from the FBI Laboratory. Thompson felt out of the loop, and he wanted Parker to fill him in. Parker agreed to come to Washington and brief Thompson personally. He called Peter Jahrling and asked him to come along.

Parker and Jahrling traveled to Washington in a green Ford Explorer driven by a sergeant wearing fatigues—this was the general's staff car. They went to the sixth floor of HHS headquarters and met with Thompson, D. A. Henderson, and other senior members of the HHS staff in a large meeting room overlooking the Mall. They were surprised to find FBI officials there, including the director, Robert S. Mueller III. Also in the room were a number of obviously powerful dark-suited officials who introduced themselves in mumbling voices. They had names like John Roberts, and they said they were from some institute or other. That is, they were top management from the CIA. Their real names were classified.

Jahrling had brought Geisbert's photographs of the anthrax particles, and he laid them out. Then he produced another something interesting for people to look at: a plastic bag containing six tubes of different orange-tan powders from the Al Hakm anthrax facility in Iraq. A friend of Jahrling's had collected them there. The powders were anthrax surrogate—fake bioweapons. A surrogate is used for testing and development of a real bioweapon. Iraqi biowarfare scientists had been making anthrax surrogate out of *Bacillus thuringensis* (BT), which is closely related to anthrax but is harmless to people. (It is anthrax for insects, and it is used by gardeners to kill grubs. The Iraqis had claimed for a while that the Al Hakm facility had been built to deal with grubs in Iraq.)

He passed the bag around the room, assuring people that the vials weren't dangerous. Everyone could see how different the Iraqi "anthrax" looked from the Daschle powder. It was heavy and crude, and contained large amounts of bentonite (a type of clay commonly used in the oil industry), and looked like lumps of dirt. It didn't look like the Daschle powder at all. At least at the time Al Hakm was running, the Iraqi bioweaponeers had been using a different formula than what was used for the Daschle powder.

Afterward, Parker suggested to Jahrling that they brief

the Pentagon on the anthrax, so they spent the rest of the day circling among the offices of assistant secretaries of defense. Toward the end of the day, they headed back up Interstate 270 to Fort Detrick. It was rush hour, and the traffic was moving like glue. Jahrling was sitting in the front seat, beside the driver, and the general was sitting in the back. On Wednesdays, Jahrling always picked up his daughter Bria at a dance class, and he was looking forward to a little bit of special time with her.

Just as the Explorer arrived at the entrance to Fort Detrick, the general's cell phone rang. The person on the other end of the line issued some rapid instructions and added, "Where's this guy Jahrling?"

"He's in the car with me." The general leaned forward to Jahrling. "We're wanted at the White House. Right now."

"Hey, General Parker—do we have time to stop and take a leak?"

"No."

The sergeant whipped a U-turn around the Abrams tank at the entrance to Fort Detrick, and they sped back onto the interstate. The sergeant started popping the lights and sirens, weaving through traffic. This wasn't helping Jahrling's state of mind. Eventually, he remembered about Bria. He called Daria and said, "I'm not getting Bria."

"What do you mean?" she asked.

"I can't tell you."

"What do you mean you can't tell me? Where are you, Peter?"

"I can't say where I am."

The car was pulling onto Constitution Avenue, and he said he'd talk to her later.

"Peter, do you still have that stuff from Iraq in your pocket?" General Parker asked. "You might not want to bring it into the White House"—the Secret Service might not react well.

They were in the White House driveway, and Jahrling didn't know what to do with his Iraqi "anthrax." He rammed it down into the crack of the car seat.

In the foyer, cabinet officials, White House staff, members of the National Security Council, senior FBI, and top-level spooks were milling around. "Where's the bathroom?" Jahrling muttered to the crowd. Someone directed him.

The meeting took place in the Roosevelt Room, which has ornate, high ceilings and oak doors decorated with brass fittings. There was a long table in the center of the room, with leather-upholstered armchairs placed around it. Many more chairs were placed around the walls.

A security official informed everyone that the meeting was secret. (The next morning, the meeting's events were described in a front-page story in *The New York Times*. White House officials later concluded that the leak had come from a source in the FBI.) Attorney General John Ashcroft sat at the table, and Robert Mueller sat close to the center, accompanied by a cluster of FBI officials, including Allyson Simons. Tommy Thompson also sat near the center of the table. The meeting was chaired by Tom Ridge, who had recently been named director of homeland security.

Jahrling started to sit on one of the chairs against the wall, but someone took him by the arm, and he was shown to a chair at the center of the table, where he faced cabinet members wearing dark charcoal suits. Jahrling was wearing his gray suit with a candy-striped shirt and a snappy necktie. The doors were closed by the Secret Service.

TOM GEISBERT had been looking for Jahrling around the Institute and couldn't find him. He got worried and called Jahrling's home, and got Daria. "Where is Peter?" she asked him. "He didn't pick up Bria!" She let Geisbert have it.

"She was as mad as a hornet," Geisbert recalled. He tried to reassure her, but he didn't know where Jahrling was either.

Daria loved Peter. It was a strong marriage, but she thought that, national crisis or not, her husband owed it to the family to at least tell them where he was.

JOHN ASHCROFT led off the meeting. He did not mince words. There was an obvious lack of communication between the Army, the FBI, and the CDC, he said, and the purpose of this meeting was to determine why the CDC hadn't realized that the anthrax was weapons-grade material and hadn't taken action faster on the Brentwood mail facility. There was a feeling that whoever had released the anthrax could do it again, perhaps with a massive release inside a landmark building or into the air of a city. This was an urgent national threat. Where did the communication break down? Had the Army given the information to the FBI? Had the FBI informed the CDC about the highly dangerous nature of the anthrax?

Ashcroft was Robert Mueller's boss, and he looked straight at the FBI director. Mueller turned his gaze to General Parker. Mueller thanked the Army for bringing the nature of the anthrax to the FBI's attention. He said that the FBI had received conflicting data on the anthrax. The FBI had been trying to sort this issue through, but Mueller now acknowledged that the Army had been right: the Daschle anthrax was a weapon.

Then twenty people around the table started arguing: what *is* a biological weapon?

John Ashcroft cut everyone off. "Okay, okay! All this discussion about what's a biological weapon is angels dancing on the head of a pin. I want to hear what the professor has to say." He pointed with his finger to some-one seated behind Jahrling.

Jahrling, who is not a professor, turned around and looked. Then he realized the attorney general meant him. Jahrling cleared his throat and directed everyone's attention to Geisbert's pictures of the anthrax skulls. (Staffers had passed them around.) He pointed out the fried-egg goop flowing off the spores in some photographs. This, he said, was probably an additive.

Someone asked, Does the professor think this anthrax could be a product of Iraq?

The best Jahrling could say was that it *could* be Iraqi anthrax, but all the samples they'd seen from Iraq, so far, were entirely different. The Iraqi anthrax had been mixed with bentonite, and these spores didn't have clay in them. He said that by tomorrow the Army would have a better idea of what the additive was.

The meeting raced off on the question of whether a "state actor" could have been behind the anthrax attacks. The atmosphere in the room started to feel like a war council deciding whether or not to attack Iraq.

Jahrling got scared. "Whoa!" he blurted. "This anthrax isn't a compelling reason to go to war. It isn't necessarily the product of a state actor." He flushed and stopped talking: saying "Whoa!" to the Cabinet seemed flippant. Then he went on. He said that a few grams of highly pure anthrax could have been made in a little laboratory with some small pieces of equipment. "This anthrax could have come from a hospital lab or from any reasonably equipped college microbiology lab." The FBI officials posed the question: how would investigators look for "signatures" of a small terrorist bioweapons lab? Jahrling answered that a small lab for making anthrax might go virtually unnoticed, and in any case would be hard to recognize.

Ashcroft closed the meeting by taking the FBI, the

Army, and the HHS to the woodshed. He gave them a stern warning to get their acts together and start communicating with one another more effectively. He made it perfectly clear that those who serve at the pleasure of the president can cease to serve in an instant.

"Well, professor, you did okay," Parker said to Jahrling on the way back to the Institute. Jahrling leaned back on the seat, and the night rushed by. He began to wonder more deeply about what he had said at the meeting—that the anthrax could have come from a small lab, a few pieces of tabletop equipment. What would it take to do the Anthrax Trick? It could be done by an individual, perhaps, or by two or three people. He started thinking about labs. There was a lab in the west. . . . There was also USAMRIID. Could that be possible? Could this be an inside job? Could it be terror coming from within the Institute? Peter Jahrling had the dizzying thought that the terror might just be coming from someone he knew or knew of.

HE GOT HOME after midnight. Daria had retrieved Bria at the dance class and had put Kira to bed. She was sitting in the kitchen grading a pile of English papers. "Where have you been? I'm sure it was somewhere important."

"I was at the White House."

"Okayyy."

"No, really."

"And you couldn't tell me."

"No, really, I couldn't."

Some days later, the general's driver stopped by Jahrling's office with the bag of Iraqi "anthrax." He said he had found it stuck in his car seat.

Tricks

KEN ALIBEK is a quiet man, in early middle age, with youthful looks. He dresses elegantly, in fine wool jackets and subdued ties. He comes from an old Kazakh family in Central Asia. Alibek arrived in the United States in 1992, through a chain of events that involved the CIA. Before then, he was Dr. Kanatjan Alibekov, the first deputy chief of research and production for the Soviet biological-weapons program, Biopreparat. Dr. Alibekov had thirty-two thousand scientists and staff working under him. When he arrived in the United States, he was overweight and depressed, and he spoke no English.

Ken Alibek has a doctor of sciences degree in anthrax. It is a kind of super-degree, which he received in 1988, at the age of thirty-seven, for directing the research team that developed the Soviet Union's most powerful weapons-grade anthrax. He did this work when he was head of the Stepnagorsk bioweapons facility, in what is now Kazakhstan; it was at one time the largest biowarfare production facility in the world. The Alibekov anthrax became "fully operational" in 1989, which means that it was loaded into bombs and missiles.

The Alibekov anthrax, as Alibek described it to me, is an amber-gray powder, finer than bath talc, with smooth, creamy, fluffy particles that tend to fly apart and vanish in the air, becoming invisible and drifting for miles. The particles have a tendency to stick in human lungs like glue. Alibekov anthrax can be manufactured by the ton, and it is believed to be extremely potent.

One day, Alibek and I were sitting in a conference room in his office in Alexandria, Virginia, and I asked him how he felt about having developed a powerful biological weapon. "It's very difficult to say if I felt a sense of excitement over this," he said. His English is perfect, though he speaks it with a Russian accent. "It wouldn't be true to say that I thought I was doing something wrong. I thought I had done something very important. The anthrax was my scientific result. My personal result."

I asked him if he'd tell me the formula for his anthrax.

"I can't say this," he answered.

"I won't publish it. I'm just curious," I said.

"You must understand, this is unbelievably serious."

Alibek gave me the formula for his anthrax in sketchy terms. The formula appears to be quite simple and is not exactly what you might expect. Two unrelated materials are mixed with pure powdered anthrax spores. If you walk into a Home Depot and look around, you may find at least one of the materials and possibly both of them. To have perfected this trick, though, must have taken plenty of research and testing, and Alibek must have driven his group with skill and determination.

"That was my contribution," he said.

WHEN KEN ALIBEK defected, his CIA debriefers discovered that they did not understand what he was talking about. Since the end of the American bioweapons program in 1969, the CIA had lost most of its expertise in biology. The Agency called in William C. Patrick III to help with the debriefings. Patrick, who is a tall, courtly, genial, balding man, now in his seventies, had been the chief of product development for the Army's biowarfare program before it was shut down in 1969. Bill Patrick holds a number of classified patents—so-called black patents—on the ways and means of making a biopowder that vanishes in the air and can drift for many miles.

Patrick and Alibek had long conversations in motel rooms, always observed and managed by handlers. The two bioweaponeers were among the very top scientists in their respective programs, and they discovered that they talked the same scientific language. As they became acquainted with each other, they found that they and their research teams had independently discovered the tricks that make biopowders fly into the air and vanish. Patrick and Alibek became friends. Patrick and his wife, Virginia, began having Alibek over for Thanksgiving and Christmas, because they felt he was lonely.

ONE DAY a few years ago, I drove up the slopes of Catoctin Mountain on a winding country road. It was a cold, raw day, and winter clouds over the mountain formed lenses that let in loose splashes of sunshine. The Patricks live in a comfortable house that resembles a Swiss chalet. It sits at the high point of a small meadow on the mountain, looking down on Fort Detrick. From the house, you can see the roof and vent stacks of USAMRIID, nestled among trees in the distance.

"Come in, come in, young man," Patrick said. He squinted up at the sky. He is exquisitely sensitive to weather.

We sat in the living room and chatted. "There's a hell of a disconnect between us fossils who know about biological weapons and the younger generation," he said. After the offensive program was closed down, Patrick joined USAMRIID for a while, doing peaceful work, but he became quite certain that one day some knowledgeable person was going to use a germ weapon in a terrorist attack, and he began a personal campaign to warn the government of the danger. He was a consultant to various agencies and governments, including the city of New York, and he gave presentations in which he described what small amounts of different powdered bioweapons would do in the air. He also gave estimates of casualties. His projections for a bioterror attack in New York City would appear to be classified.

A few minutes after I arrived, Ken Alibek showed up, driving a silver BMW. After lunch, we settled around the kitchen table. Patrick brought out a bottle of Glenmorangie single-malt whisky, and we poured ourselves drams. The whisky was golden and warm, and it moved the talk forward.

"There seems to be a belief among many scientists that biological weapons don't work," I said. "You hear these views quoted a lot."

The two ex-bioweaponeers looked at each other, and Bill Patrick let out a belly laugh, put his head down, and kept on laughing. Ken Alibek looked annoyed. "This is so stupid," Alibek said. "I can't even find a word to describe this. You test the weapons to find out what works. I can say I don't believe that nuclear weapons work. Nuclear weapons destroy everything. Biological weapons are more . . . beneficial. They don't destroy buildings, they only destroy vital activity."

"Vital activity?"

"People," he said.

Patrick invited us into his basement office. We followed him down a spiral staircase to a room that had sliding glass doors. He took a paper bag out of a filing cabinet, and he pulled out a little brown glass bottle. The bottle had a black plastic cap that was screwed on tightly, and it was half full of a cream-colored, ultrafine powder. "That's a simulant anthrax weapon," he said. "It's BG"—*Bacillus globigii*, a harmless organism related to anthrax. "Take a look at that, Ken."

Alibek held the bottle up and shook it. The powder turned into a cloud of smoke inside the bottle. The smoke swirled around, and the bottle went opaque.

"Now, that is a beautiful product," Patrick remarked.

Alibek nodded. "It has the characteristics of a weapon."

Patrick removed an insecticide sprayer from the paper bag. It was an old-fashioned hand-pump flit gun. He pumped the handle, and a cloud of white smoky powder boiled out of the nozzle. "Isn't that a beautiful particle size?"

Alibek started laughing. "Don't point that thing at me, Bill!"

"It's actually my wife's bath powder." A pleasant scent of baby powder filled the room.

The room had become a bit stuffy with the powder, so we went outdoors on the lawn in front of the house. Alibek lit a cigarette, and we admired the view down the meadow and over the piedmont of Maryland to a blue line in the distance, the Mount Airy ridge. The patchy clouds now covered the sun.

"Wind's ten to twelve miles an hour, gusting a bit," Patrick said. "Which way is the wind going, Ken?"

Alibek turned around and looked up. He seemed to be feeling the air with his face. "East? It's going east."

"Smallpox would get to Frederick from here on a day like today," Patrick remarked.

Alibek nodded in agreement and pulled on his cigarette.

"Hold on," Patrick said abruptly, and he strode up the

hill and disappeared around the corner of the garage. We heard the electric motor of the garage door. He returned in a few moments, carrying a mayonnaise jar that contained a powder. He unscrewed the metal lid and showed me the jar's contents. It was half full of an extremely fine powder of a mottled, pinkish color. He explained that it, too, was a simulated bioweapon. The pink color in the powder came from the blood of chicken embryos. The powder was a surrogate of a weaponized brain virus called VEE, which travels easily in the air—but the powder was sterile and had no infectious material in it. He shook the jar under my face, and smoky, hazy tendrils wafted toward my nose. I fought an urge to jerk my head back—the mind may know the fog is harmless, but the instincts are hard to convince.

Patrick walked across the lawn with the jar and stood by an oak tree. Suddenly, he straightened his arm and heaved the contents of the jar into the air. The powder boiled out, making a small mushroom cloud, and then the simulated brain virus blasted through the branches of a dogwood tree and took off down the meadow, moving at a fast clip toward Frederick. Within seconds, the cloud started becoming transparent, and then, abruptly, it vanished. The particles seemed to be gone. It had looked like steam coming out of a teapot.

"See how it disappears instantly?" Patrick remarked.

Alibek watched, tugging at his cigarette, mildly amused. "Yeah. You won't see the cloud now," he said. "Depending on the altitude of the dispersal, some of those particles will go fifty miles."

"Some of them'll get to the Mount Airy ridge. It's twenty miles away," Patrick said. The simulated brain weapon would arrive at the ridge in a couple of hours. A couple of hours after that, the simulated brain virus would be beyond the horizon.

Patrick was eyeing the clouds, seeming to sniff the wind. He turned to Alibek. "Say you wanted to hit Frederick today, Ken, what would you use?"

Alibek glanced at the sky, weighing the weather and his options. "I'd use anthrax mixed with smallpox."

Stew Phone

OCTOBER 25, 2001

TOM GEISBERT drove his beat-up station wagon to the Armed Forces Institute of Pathology, in Northwest Washington, carrying a whiff of sterilized dry Daschle anthrax mounted in a special cassette. He spent the day with a group of technicians running tests with an X-ray machine to find out if the powder contained any metals or elements. By lunchtime, the machine had shown that there were two extra elements in the spores: silicon and oxygen.

Silicon oxide.

Silicon dioxide is glass.

The anthrax terrorist or terrorists had put powdered glass, or silica, into the anthrax. The silica was powdered so finely that under Geisbert's electron microscope it had looked like fried-egg gunk dripping off the spores.

Geisbert called Jahrling on an open telephone line and said, "We have a signature of something." Jahrling asked him to stop talking on an open line.

Geisbert asked someone if he could use the stew phone, and he was shown into a secure room. The stew phone looked like a normal telephone, except that it had an LCD screen and an encryption lock. They gave Geisbert the encryption key, and he unlocked the phone.

Jahrling, meanwhile, had gone to the Secure Room at USAMRIID. He unlocked his stew phone and waited. Geisbert called in, they spoke a few words in open mode, and then Jahrling pushed a button on the phone. The screen flashed: GOING SECURE.

The phones went silent. The two men waited half a minute. Then the screen on the stew phone read: US GOVERNMENT SECRET, and their voices came back on the line, distorted.

"So—what—do—you—have?" Jahrling said.

"Wisten, Weet! We ow-wowo-wooow, wow." Geisbert's voice turned into a stretched-out robo-gargle.

"Slow—it—up."

"We fow wow-wow!"

"Whoa. You—have—to—speak—distinctly."

"Pete! There's—glass—in—the—anthrax."

YOU COULD GO ON the Internet and find places to buy superfine powdered glass, known as silica nanopowder, which has industrial uses. The grains of this type of glass are very small. If an anthrax spore was an orange, then these particles of glass would be grains of sand clinging to the orange. The glass was slippery and smooth, and it may have been treated so that it would repel water. It caused the spores to crumble apart, to pass more easily through the holes in the envelopes, and fly everywhere, filling the Hart Senate Office Building and the Brentwood and Hamilton mail-sorting facilities like a gas.

No one knows how many anthrax spores leaked into the air at the Brentwood mail facility. At least two letters containing dry skull anthrax went through the machines. The skulls were crumbling and falling apart, and individual spores were leaking through pores in the paper and perhaps coming out through the corners of the letters. If all of the spores that went into the air inside the Brentwood building were gathered into a heap, it's doubtful they would have covered the head of a thumbtack. The Environmental Protection Agency spent an estimated thirty million dollars trying to get rid of the spores there.

The Feds

THE WASHINGTON FIELD OFFICE of the FBI is a new stone-and-glass building at Fourth and F streets, a few blocks east of the FBI headquarters, on the edge of Chinatown. The Washington office was given overall management of the criminal investigation into the anthrax attacks, which came to be called Amerithrax. There were five homicides in the Amerithrax case. Robert Stevens in Boca Raton and the two Brentwood postal workers, Joseph Curseen, Jr., and Thomas Morris, Jr., were the first to die. Then a sixty-one-year-old woman in New York City named Kathy Nguyen became ill and died of inhalation anthrax; the source of her exposure was never identified. On the day before Thanksgiving, in Connecticut, a ninety-four-year-old woman named Ottilie Lundgren also died of anthrax. The source of her exposure was not found either, but was likely to have been a few spores that she inhaled from a piece of mail that had touched some other piece of mail that had gone through the Hamilton, New Jersey, sorting facility and had probably been in close contact with an anthrax letter. This was a murder and terrorism case that cut across jurisdictions. The FBI termed it Major Case 184.

The Washington field office was run by an assistant director of the FBI named Van A. Harp. Directly under him were three special agents in charge of the office, or SACs. One of the SACs was Arthur Eberhart, who had served earlier as a section chief at Quantico, overseeing the Hazardous Materials Response Unit. In early October, as the first anthrax deaths occurred, Eberhart began assembling assets—calling people into the team, sometimes drafting them out of other units, "for the needs of the Bureau." A working group formed up quickly, and eventually it became two squads, known as Amerithrax 1 and Amerithrax 2. Eberhart put John "Jack" Hess in charge of Amerithrax 1 and David Wilson in charge of Amerithrax 2. Hess's squad handled much of the classic detective work, while Wilson's squad took care of the scientific side of the investigation. Jack Hess and David Wilson were basically given the job of solving the Amerithrax case.

I first met David Wilson in 1996, when I was doing some research at the FBI Academy at Quantico, and he had just been assigned to the HMRU as an agent. He was a quiet man who stayed in the background and said little, but like many FBI people, he had a casually aware manner, as if there was a part of him that was always evaluating things. At that time, FBI scientists were saying that a bioterror attack could be very difficult to solve, because the evidence left in its wake might only be dead people with a strain of a microorganism in their bodies, and precious little else. One evening, I drank beers with some FBI scientists at the Quantico Boardroom, a bare-bones cafeteria and pub, and they started tossing out all sorts of ideas about how you would actually solve a bioterror crime. Most of them were high-tech solutions, involving sensor machines and exotic lab techniques, but a section chief named Randall Murch, who had created the Hazardous Materials Response Unit, told the group that he thought that, in the end, traditional detective work would solve a biological crime. "Ultimately, humans make mistakes," Murch said.

DAVID LEE WILSON is a tall man in his mid-forties, with broad shoulders and large hands. He has straight brown hair, dark eyebrows, and pale gray eyes. On the job, he usually wears a starched white button-down shirt. He was raised in Tennessee, in a farmhouse that his grandfather built out of sawn planks of poplar, and he has a Tennessee accent. When he speaks, his voice goes along rapidly and softly over a wide range of topics. He has a degree in botany, with an emphasis on marine biology. He spent time on research ships studying the biological productivity of seas full of phytoplankton. When he joined the FBI, he gravitated to the forensic examination of trace evidence. At home, to relax, he picks a Martin acoustic guitar. He picks precisely and with a flowing musical sense. He told me that he doesn't like attention. "It makes me uncomfortable to have any kind of single focus on me," he said. He was careful to explain to me that he was only one member of a large FBI operation. "Teamwork is critical for this case," he said. "A major case is like an organism. It is almost alive. It changes in response to evidence that comes in, and it has feedback loops."

Wilson was the head of the HMRU between 1997 and 2000, and during those years the number of credible bioterror threats or incidents rose dramatically, up to roughly two hundred a year, or one biological threat every couple of days. Most of them were anthrax hoaxes. The HMRU teams were constantly doing flyaways, taking helicopters or FBI fixed-wing aircraft to various places around the United States in order to assess a threat of anthrax and collect evidence. Running the HMRU was a little like running a firehouse that went out on a lot of false alarms, and Wilson got a little tired of it, particularly because he was trying to build a national program and kept finding himself sitting on a jump seat in a Huey loaded with biohazard equipment, flying to another bioscare. His young daughter would ask her father to leave his cell phone behind when they went to a restaurant, and if his pager beeped, she would roll her eyes and say, "Not again, Daddy." Wilson wanted to supervise field investigations in which he could develop and pursue criminal cases. He ended up transferring to the Washington Field Office. Then along came Amerithrax, and they put him in charge of the science in the case.

Wilson's case strategy for Amerithrax 2 involved reaching out across the spectrum of scientific talent in the United States and getting help wherever he could find it. He developed relationships with the national laboratories (which are run by the Department of Energy), with the Defense Department, the CIA, and with the National Academy of Sciences and the National Science Foundation. He recruited dozens of outside scientists—chemists, biologists, geneticists. He pulled in a Navy expert in anthrax named James Burans, and he took in a CDC epidemiologist, Dr. Cindy Friedman, who joined Amerithrax 2 as a full-time squad member.

Kenneth C. Kohl, an assistant U.S. attorney, was attached to the Amerithrax squads full-time, and he moved into an office in the building on Fourth and F streets. He advised agents about developing evidence that could be used in court. The FBI was mindful of the case of Richard Jewell, a security guard whom the FBI had suspected of planting a bomb in Centennial Park in Atlanta during the summer Olympics in 1996. Jewell was exonerated, and it was a huge embarrassment to the FBI; it made the Bureau look incompetent and prejudiced, and the case is still unsolved. Of all the pressures hitting the Amerithrax agents, the most potent was the knowledge that, in the end, all the paths of Amerithrax led to a jury.

It was quite possible that if anyone was charged with the Amerithrax crimes, Kohl might seek the federal death penalty. But to bring a prosecution in a multiple murder case in which the murder weapon was a living microbe, the evidence would have to be tight and clear, persuasive to a jury, and sharp with proof—probatory, in the language of police work. There would not necessarily be any testimony from eyewitnesses. The crimes could have been perpetrated by one person acting alone, and so the Amerithrax case might have to be tried largely on forensic evidence: on the science squad's work. "I wonder, though, if Randy Murch's words of yesteryear may prove prophetic for Amerithrax," Wilson said, recalling that evening in the Quantico Boardroom. "We just don't know how it's going to go, and sometimes you just get lucky. Somebody calls you and says, 'You know, I saw something.' And you say to yourself, *'That's it.'*"

Amerithrax became one of the most complex cases ever run by the FBI. The two Amerithrax squads occupied half of the seventh floor of the Washington field office. Each squad was small, with only about ten or so members, but they were supported by teams of analysts, and the squads were given the power to order practically anyone in the FBI to follow a lead or accomplish a task. There are twenty-five thousand people in the FBI. The Amerithrax squads used them to cover thousands of leads, and they relied on the work of many other people across the federal government.

Trenton was an obvious place to examine, and FBI agents went all over the area, looking for sites where the letters had been mailed, setting up surveillance, checking out connections to possible al-Qaeda suspects. But there was remarkably little to go on. Wilson and his squad began grinding on the science of the case. "Not that Dave won't work the case to death," a former top FBI official said to me, "but basically all the leads, all you get, are what is captured in the biological material in the letters, in the tape that sealed the letters, and in the writing in the letter itself."

The Quantico behavioral profilers went to work on the handwriting and language of the letters. The profilers came to be convinced that the anthrax terrorist was a white male, a loner, perhaps quite shy, with a grudge, and with scientific training, and they felt the terrorist would be a native speaker of English, not Arabic. A native speaker of Arabic would be more likely to have written "God is great," not "Allah is great."

ON NOVEMBER 16th, another anthrax-laden letter was found in a sealed plastic bag full of mail. This letter was addressed to Senator Patrick Leahy of Vermont. It was among the mail in the Hart building that had been sequestered. The Leahy letter contained something like a gram of finely powdered anthrax spores, bone white, treated the same way as the Daschle spores. The FBI delivered the Leahy letter to USAMRIID, where diagnostic scientists began analyzing the powder.

FBI forensic experts in hair and fiber analysis also examined the letter, most particularly the tape that sealed the envelope. Tape is a valuable forensic material because it picks up dust, including tiny fibers of hair, carpet, and clothing. Forensic samples that are collected from criminal evidence are known as questioned samples, or Q samples, because they come from an unknown ("questioned") source—which may be associated with the unidentified perpetrator of the crime. These Q samples may be matched to known samples, or K samples, which are reference samples that are fully identified. In this way, trace evidence can be understood and can be linked to a known source, such as the perpetrator or the perpetrator's environment. A single human hair can contain unknown human DNA—a questioned sample of DNA—which can be matched to a known sample of a person's DNA. The FBI's hair and fiber experts can take a particular questioned fiber and match it precisely to a fiber that has come from a known manufacturer in a particular color and style. Manufacturers use constantly changing formulas for dye and for materials, and fibers can come in all sorts of sizes and shapes—round, delta, trilobal, oval, wrinkly. The top hair-and-fiber person in the FBI is a unit chief named Douglas Deedrick, who works at the Laboratory at FBI headquarters. They say that Deedrick has a near photographic memory for fibers he may have seen just once before in his career. He'll throw out a line of patter: "I've seen this before. . . . I know this fiber. . . . That's a carpet fiber from a stinkin' seventy-three Bonneville," is the sort of thing he can say when he's working. If a Q sample can be matched to a K sample, it can have probative value—it can lead to a suspect and, ultimately, to a conviction in a criminal trial. (When O.J. Simpson struggled to put on the glove at his murder trial, he gave a dramatic show to the jury of an apparent blundering attempt by the prosecution to try to match something questioned to something known—the glove to his hand.)

The FBI's forensic scientists apparently had great difficulty getting Q samples from the letters. They won't comment, but it seems that they found no hairs or fibers of particular interest on the tape. The anthrax terrorist or terrorists had perhaps been quite careful to load the letters in an environment that was free of dust and hair—possibly inside a laminar flow hood. They did find that the cut edges of the strips of tape matched one another. The perpetrator had loaded and taped the envelopes one after the other using the same roll of tape. They tested the paper of the envelopes for human DNA, using the PCR (polymerase chain reaction) method, which can amplify tiny trace amounts of DNA. The method is so sensitive that if a person breathes on a sheet of paper, the paper can retain fragments of the person's DNA that can be detected. There was apparently no questioned human DNA found on the envelopes or on the stamps. This suggested that the perpetrators might have worn a breathing mask while loading the letters. There were no questioned fingerprints on the letters, either, which probably meant that the perpetrators had worn rubber gloves. The anthrax terrorist or terrorists seemed to have been careful to avoid leaving any evidence on any of the letters. What was left was the powder inside the envelopes, and the handwriting and contents of the letters. Those were apparently the best Q samples that the FBI had to go on, and it was precious little.

In November, the microbiologist Paul Keim, working with his group at Arizona State University in Flagstaff, identified the strain in all the anthrax letters as the Ames strain. It had been collected from a dead cow in Texas in 1981, and had ended up in the labs at USAMRIID.

USAMRIID scientists had later distributed the Ames strain to a number of other laboratories around the world. By showing that the strain in the letters was the Ames strain, Paul Keim gave the FBI a sort of incomplete or partial K sample: it was not a really precise K sample, but further analysis of the strain in the letters might provide a tighter match to some known substrain of the Ames anthrax. The Ames strain was natural anthrax. It had not been "heated up" in the lab—had not been genetically engineered to be resistant to antibiotics. Nowadays it is so easy to make a hot strain of anthrax that's resistant to drugs, intelligence people simply assume that all military strains of anthrax are drug resistant. The fact that the Amerithrax strain wasn't military pointed to a home-grown American terrorist rather than to a foreign source, to someone who had perhaps not wanted large numbers of people to die. Someone who might have wanted to get attention.

THE CIA had a secret program called Bacchus, in which a group of researchers with the Science Applications International Corporation (SAIC), working at the U.S. Army's Dugway Proving Ground in Utah, built a miniature anthrax bioproduction plant using inexpensive, off-the-shelf equipment. The idea of the experiment was to see if it would be possible for terrorists to buy ordinary equipment, make anthrax with it, and not be noticed. In January and February 2001, roughly ten months before the anthrax terror event, the Bacchus team succeeded in making a powdered anthrax surrogate, BT, but it was crude. Now the FBI investigators focused a lot of attention on scientists who had access to Dugway, where the U.S. military tests various biosensor systems and where there are stocks of anthrax.

The Amerithrax squads seemed to have a case that was cooling off. The FBI was letting it be known—whether accurately or not—that the list of potential suspects had never gone below about eight individuals and was really more like twenty to thirty people.

There were mysteries and loose ends that seemed to baffle the FBI, including hints that the anthrax might have been part of an al-Qaeda terror operation. In January 2001, several of the men who would later hijack the four airplanes involved in the September 11th attacks rented apartments near Boca Raton, Florida. The real-estate agent the men dealt with was the wife of an editor at American Media, where Robert Stevens, the first man to die of the anthrax, worked—but the real-estate agent felt that the hijackers could not possibly have known that her husband worked there. Mohammad Atta, who was believed to be the operational leader of the hijackers, made inquiries at airports in Florida about renting crop-dusting airplanes: he obviously had it in mind to spray something from the air. In June 2001, two men, Ahmed al-Haznawi and Ziad al-Jarrah, who would later be among the hijackers of United Flight 93, which crashed in Pennsylvania, went to the emergency room of the Holy Cross Hospital in Fort Lauderdale. Al-Haznawi was complaining of an infection on his leg, and an emergency-room doctor named Christos Tsonas examined him. The man had a blackened sore on his leg that he told Dr. Tsonas he had gotten from bumping into a suitcase. The doctor didn't think that sounded likely. He prescribed antibiotics to al-Haznawi and never heard from the men again. Tsonas contacted the FBI in October and told agents that the sore had been consistent with cutaneous anthrax. Agents apparently went through the hijacker's possessions and swabbed them for anthrax spores, and found none. "We've debated that one informally a lot around our shop," an FBI source at Quantico told me. "Everything I've heard basically discounts it."

In Trenton, FBI investigators became interested in various people living in an apartment complex called Greenwood Village. They arrested a man, Mohammad Aslam Pervez, who was listed in the phone book as living there. Pervez was thirty-seven years old, a naturalized American citizen born in Pakistan, and he had recently worked at a newsstand in the Trenton train station and also at a newsstand in the Newark train station with Mohammad Jaweed Azmath and Ayub Ali Khan, who were arrested on September 12th on an Amtrak train in Fort Worth, Texas, carrying box cutters, five thousand dollars in cash, and hair dye. The FBI evidently suspected that they were al-Qaeda hijackers who had not been able to get on a plane. Pervez had lived with them in an apartment in Jersey City, while listing his address as Greenwood Village, and he was allegedly moving large amounts of money around. The FBI charged Pervez with lying to federal investigators about the nature of more than $110,000 in checks and money orders. The neighbors in Greenwood Village told reporters that they had noticed unusual numbers of Arabic-speaking men congregating in Pervez's apartment during the summer, in the months before September 11th. A reporter from *The Wall Street Journal* managed to get inside the Jersey City apartment, where he found articles clipped from *Time* and *Newsweek* on the use of sarin nerve gas and biowarfare agents. On October 29th, FBI agents raided another apartment at Greenwood Village. Eight to ten agents carted away many trash bags full of evidence. An FBI spokesperson, Sandra Carroll, told reporters that the September 11th and anthrax investigations were "not necessarily separate."

But it just didn't seem to go anywhere.

A few months before the first anniversary of the

anthrax attacks, I visited the Amerithrax squads in the Washington field office. The two squad supervisors, Jack Hess and David Wilson, had offices side by side, facing an open floor of cubicles. The CDC doctor on the squad, Cindy Friedman, was meeting with two FBI agents, talking about something in low voices. They asked me to stand out of hearing when there were any discussions about the case. Large posterboards leaned against filing cabinets, covered up from view.

David Wilson led me to his office, where we ate salads from the FBI canteen for lunch. The Capitol's dome and the top of the Hart Senate Office Building were visible from the window. His office was almost bare. Three heavy briefcases sat on a desktop, and a table had a full in-box. "Until we have someone under arrest and charged with a crime, we literally can't rule anything out," he said to me. The Amerithrax case held many dimensions of crime, but at bottom it was murder. "I don't give a rat's tail for what they thought they were doing when they mailed the letters. People died," Wilson said. "Damaged facilities can either be repaired or replaced. The Brentwood building can be fixed. But the deaths can't be fixed."

ONE DAY, I spoke with a scientist who is an expert in forensic evidence, knows a lot about biology, and until recently was an influential executive in the FBI. "The Unabomber took seventeen years to solve," he said. "We just don't know who these perpetrators are, and it could be years before we get a break. I'm saying 'they.' I personally find it hard to believe that it was done by only one person. That's just gut. I don't know why, I can't put my finger on it, but if I wanted to keep tight operational security I would send a package of anthrax to someone else with instructions for how to load the envelope and mail it—you know, 'Don't lick the envelope, do this, do that.' I would do it with opsec."

"Opsec?"

"Opsec—operational security. It's a standard security approach for making yourself as invisible as possible. There's a leader who organizes and directs an operation, and a different person carries it out." The person who does the operation is expendable. The September 11th attacks were done with opsec, and the Palestinian suicide bombings feature opsec. He went on: "I have a feeling that, in the end, it's going to be like one of our fugitive cases, where a girlfriend rats on the guy or someone talks. I'm a forensic scientist, but unfortunately I have a feeling that traditional investigation is going to solve this case in the end, not science."

Ebola in the Afternoon

BARBARA HATCH ROSENBERG, the chair of the Federation of American Scientists' Biological Arms Control Program and a professor of environmental science at the State University of New York at Purchase, believed that the anthrax terrorist was an American scientist. She began speculating, in speeches and on a website, that the terrorist was a white male who had worked in classified programs for the government. She wondered publicly if the terrorist had once worked for USAMRIID or another government laboratory. She felt that the terrorist might have been a contractor working for the CIA, with access to secret information about government involvement with offensive biowarfare programs. Rosenberg is a trim, middle-aged woman with a forceful manner, and she is not afraid to speak her mind. Her web site got a lot of traffic, and in late June 2002, Senators Tom Daschle and Patrick Leahy asked her to come meet with them. She was very happy to oblige them.

A few days later, the FBI searched the apartment of Dr. Steven Hatfill, in Frederick. Hatfill, the colorful Ebola researcher who had trained Lisa Hensley in blue-suit work and who liked to eat candy bars in his space suit, had left USAMRIID in 1999 and gone to work for Science Applications International Corporation, the defense contractor that conducted the CIA's Bacchus program. Hatfill was divorced and had continued to live in Frederick after he left USAMRIID. He lived by himself in the Detrick Plaza apartments, a brick complex right next to the gate of Fort Detrick, a stone's throw from the Abrams tank. From his apartment unit, he could look over a fence and across a lawn, where he could see the FBI helicopters coming and going next to USAMRIID, ferrying evidence from the Amerithrax case. FBI agents arrived at Hatfill's apartment with a rented Ryder truck. (The apartment manager told a reporter that Hatfill was "traveling abroad" when the FBI came.) They put on bioprotective suits and searched the apartment, and then removed some computer devices and plastic bags of Hatfill's possessions, which they loaded into the truck and took away. Hatfill had consented to the search. He had a storage facility in Ocala, Florida, two hundred and fifty miles from Boca Raton. He also had access to a cabin in a remote part of Maryland. It was reported that he had asked visitors to take Cipro before entering it. The storage facility was not far from a ranch in Ocala called Mekamy Oaks, where Hatfill's parents, Norman and Shirley Hatfill, raised Thoroughbred horses.

The FBI said that Steve Hatfill was not a suspect in the case. He told journalists that he was cooperating with the authorities in an effort to clear his name, and he insisted that he had absolutely no involvement with the anthrax attacks. In February 2002, Scott Shane, a reporter for the Baltimore *Sun,* became interested in Hatfill. Shane spoke with Hatfill on the phone and asked him some questions, and then talked with some people who knew Hatfill. A month later, Hatfill lost his job at SAIC. Soon afterward, he telephoned the Baltimore *Sun* and left a message with the paper's ombudsman. "I've been in this field for a number of years, working until three o'clock in the morning, trying to counter this type of weapon of mass destruction, and, sir, my career is over at this time," he said. The FBI interviewed Hatfill several times, but there was nothing particularly unusual in this; the Amerithrax investigators had interviewed a number of American scientists more than once. FBI agents gave a polygraph test to Tom Geisbert.

Nonetheless, Hatfill's background attracted investigators' attention. "The Bacchus program suffered from a lack of adult supervision," a scientist said to me. (It didn't, however, produce anthrax that was anywhere near as pure as the Daschle anthrax.) Hatfill had a secret-level security clearance, and he knew Ken Alibek and Bill Patrick. Soon after he went to work at SAIC, Hatfill and a colleague commissioned Patrick to write a study on the effects of anthrax mailed in letters. Patrick, who had done many studies of this sort for the government, worked out a scenario in which a letter containing two grams of dry anthrax spores was opened inside an office building. The anthrax in Patrick's study was pure spores. Bill Patrick had imagined key elements of the Amerithrax attacks at the request of SAIC and Steve Hatfill.

Hatfill's résumé said that he had served with the Rhodesian Special Air Squadron (SAS) and with the Selous Scouts, the white antiguerrilla forces. In 1979 and 1980, during and after the Rhodesian civil war, an anthrax outbreak occurred in livestock in Rhodesia that killed large numbers of cattle, gave ten thousand people cutaneous anthrax, and killed a hundred and eighty people. The U.S. government was said to have had suspicions, and perhaps evidence, that this anthrax outbreak might have been an act of biowarfare caused by the SAS or by agents working for the clandestine South African internal-security service, the Civil Co-operation Board (the CCB). During those years, CCB people had been using biowarfare agents for assassination attempts. When he was studying medicine in Zimbabwe, Hatfill had reportedly lived a few miles from a neighborhood called Greendale. The return address of the letter to Senator Daschle was the fourth grade of the Greendale School.

After the FBI searched his apartment again, this time with a criminal search warrant, one of Hatfill's lawyers, Victor Glasberg, wrote an angry letter to Kenneth Kohl, the assistant U.S. attorney working with Amerithrax, saying that "improper decisions" had been made in the FBI's treatment of Hatfill, and that Hatfill was doing everything he could to cooperate fully with the FBI. He said he was "working with Dr. Hatfill on how to address a flurry of defamatory publicity about him which has appeared in the press, on TV, and on the Internet." Shortly afterward, Steve Hatfill read a statement to the press in front of his lawyer's office, in which he forcefully defended himself, and said he was a loyal American who loves his country, and he assailed "calculated leaks to the media" concerning him. "Does any of this get us to the anthrax killers?" he said. "If I am a subject of interest, I'm also a human being. I have a life. I have, or I had, a job. I need to earn a living. I have a family, and until recently, I had a reputation, a career, and a bright professional future."

I BECAME ACQUAINTED WITH Dr. Steven Hatfill and interviewed him in 1999, a few months before he left USAMRIID. He worked in the virology division, and he was closely connected with Peter Jahrling's research group. He was doing research in Ebola and monkeypox. Hatfill had a tiny office, with no windows, white walls, and little in the way of decoration, but he filled the room with his physical and intellectual presence. He was a vital, engaging man, with a sharp mind and a sense of humor. He was forty-five years old, with a good-looking face, brown hair, and a neatly trimmed brown mustache. He was heavy-set but looked fit, and he had dark blue eyes. I sat on top of a counter in a corner of the room, and he sat in the center of the room, in a chair at his desk, leaning back and looking up at me, and he told me a little about his life.

"I was in the Army for twenty years," he said. "I was a captain in the U.S. Special Forces, and I was in Rhodesia—Zim—but I can't say what I was doing there. I went to medical school in Rhodesia and graduated in 1984. I have two C.V.s, the classified one and the unclassified one. I've seen a lot of diseases. There was an outbreak of anthrax in Rhodesia when I was there." He went on to say that the South African CCB had been blamed for the anthrax, but he didn't think it was likely. "It was not a weapon. It was a natural outbreak that happened because there was a harsh terrorist war going on and a breakdown of veterinary health."

He was having a great time doing research at USAMRIID.

"Where else can you work with monkeypox in the morning and Ebola in the afternoon?" he said. He explained that he was working to develop antiviral drugs for smallpox. His quest was similar to Lisa Hensley's and Peter Jahrling's. Like them, he regarded smallpox as the number one threat. He wanted to find some way to test and develop drugs that would work on smallpox. He had an idea that smallpox and drugs could be tested directly on human tissue with the help of machines.

Hatfill's office had small pieces of equipment sitting in it, of types that I did not recognize. Hatfill was a gadgeteer. He picked up a glass cylinder about the size of a soda can, with metal ends, and handed it to me. "Take a look at that."

I held it, but I had no idea what it was.

"It's a bioreactor. It's called an STLV. It was developed at NASA. You can grow human tissues in it and then infect them." He explained that using a device of this sort, you could test new drugs against smallpox and other exotic diseases that could not be tested ethically in people. In other words, you didn't necessarily have to test smallpox in animals—you might be able to test the virus in a machine. He was optimistic that there would be drugs to cure smallpox, and he felt that machines would speed up the discoveries. "You can put a bit of tonsil tissue in this thing, and it actually grows a tonsil," he said.

"The bioreactor grows a tonsil?"

He grinned. "You get a tonsil. The architecture of the tissue is preserved."

"Could you grow a finger?"

Hatfill started laughing, and explained that someday we might actually be growing spare body parts in bioreactors. He explained how it worked. "What you do is, you collect tissue from the body, and you chop it up. You can use prostate tissue, lung-cancer tissue, liver, lymph, spleen.

You put the tissue pieces in the reactor, and you fill it with growth medium. The bioreactor turns around on a motor." He demonstrated by turning the device in his hands. "As it turns around, you get excellent perfusion of the tissues, and the blood vessels start to go everywhere. Then you add Ebola, and then you can do tests of drugs. I've got four of these units running in BL-4 right now." He added that he had another device in the hot lab that looked like "something out of *Star Trek*." He was using it to run tests on monkey blood infected with monkeypox.

Hatfill felt strongly that a bioterror event could happen one day, and he feared it could be very bad. He took me down the hall to see a bioterrorism-emergency storeroom. The room was full of racks holding boxes of safety gear and face masks and portable Racal space suits. "If there's an attack on a city with a large area of coverage," he said, "one third of the population will try to flee, and so you won't be able to get into the city by road. We can stockpile emergency supplies on trains. The system we envision has twenty-seven trains, to address what to do with twenty thousand casualties. Do you know what this is?" He showed me another gadget, a big one, a kind of motor with tubes, sitting next to a biohazard stretcher. "That's a mobile embalming pump." He explained that USAMRIID's emergency planners kept one on hand for disinfecting the bodies of the victims. "Once you've got the formalin in you, you're no longer infective, and we can give you some semblance of a Judeo-Christian burial," he said.

SOME OF STEVEN HATFILL'S CLAIMS about himself didn't check out: the Army said he had not served in the U.S. Special Forces. On at least one of his résumés, he had claimed to have a Ph.D. in cell biology from Rhodes

University in South Africa; officials there insisted he had never been awarded a Ph.D from that institution. He was given a secret-level national-security clearance in 1999, around the time he went to work for SAIC. Then, in 2001, he had applied for a higher-level clearance, and so he was given another background check. The government suddenly removed all of his security clearances in August of 2001, two months before the anthrax letters were mailed.

Microbiologists are naturalists, and like naturalists everywhere, they like to collect examples of interesting creatures. They can amass large and varied collections of microscopic life-forms, and often they have their own freezers and their own private labeling systems for vials. When a researcher retires, dies, or moves on, his or her freezer typically hangs around. As long as a freezer is plugged in and running, whatever is inside it will continue to exist. Once the freezer's owner is gone, the freezer can just sit there unnoticed, a mystery freezer. One day in August 2002, somebody noticed such a freezer sitting in hot suite AA5 at USAMRIID—the Ebola suite. The freezer had been used by Dr. Steven Hatfill when he worked as a postdoc there. It contained many vials and samples of pathogens with which Hatfill had been working. An FBI HMRU team put on space suits, entered AA5, opened Steve Hatfill's freezer, and removed all of its contents—a large number of vials. The vials were taken out of the hot zone and transported inside a sealed biohazard container to the CDC, where Hatfill's collection was placed in the Maximum Containment Lab. The frozen contents of the vials was being thawed and was undergoing analysis by investigators in the final months of 2002.

Aftermath of an Experiment

DURING THE ANTHRAX EVENT, Lisa Hensley kept her head down and worked on her smallpox data. Nobody from the FBI called her or gave her a polygraph exam, and she felt oddly disappointed about that. She was not involved with the anthrax investigation at USAMRIID. Meanwhile, the scientific community had begun to hear rumors that Peter Jahrling and his team had re-created smallpox in monkeys and that Jahrling had plans to write a paper about it. D. A. Henderson, who was now working inside the U.S. government, was clearly not happy about this monkey work, but he couldn't speak out in public because the official policy of the government was to develop alternatives to the traditional vaccine.

Henderson felt that a stockpile of the traditional vaccine would be more than adequate. He worked with officials from the CDC to develop a national plan for a smallpox emergency. The CDC would give ring vaccinations to the affected populations, and if those failed, everyone who could tolerate the vaccine would get it. At the same time, the U.S. Public Health Service (the parent of the CDC) would institute quarantines around cities. The National Guard would most likely have to be involved, and so the plan had elements of martial law.

• • •

WHEN HENDERSON had retired as the dean of the Johns Hopkins School of Public Health, he was replaced by an epidemiologist, Dr. Alfred Sommer, who had worked in the CDC's Epidemic Intelligence Service during the years of the Eradication. In 1970, when the cyclone hit Bhola Island, which would inspire Larry Brilliant and Wavy Gravy to go there to try to help, Al Sommer was already there. He happened to be stationed in Bangladesh with the CDC, and he ended up organizing help in an area of jungle islands in the Ganges Delta known as the Sunderbunds, not far from Bhola Island. He pioneered some of the first techniques of disaster-assessment epidemiology, methods that are now used everywhere to monitor diseases in populations that have been hit with a natural disaster.

Not long afterward, Bangladesh won its independence from Pakistan. During the civil war, ten million refugees ended up living in camps just inside India, where smallpox broke out. Sommer fought smallpox for two months in the refugee camps, often the only medical doctor at the scene. "It was just me and a couple thousand cases of smallpox, which meant five hundred to eight hundred deaths," he said. He discovered that local cemeteries were a good place to trace the movement of the virus. "People buried their dead in Bangladesh rather than cremating them, as they do in India," he said, "and they always knew when a person died of smallpox." He studied the registries of burials, and he could see the rising and falling of the generations of the virus. He used this information to determine where to set up a ring, where to vaccinate people. Today, Sommer keeps a certificate from the WHO on the wall of his office, noting his participation in the Eradication. He is as proud of it as of his Lasker Award, which is the most prestigious award in medicine. He received the Lasker for research in vitamin A deficiencies and blindness.

One day in January 2002, Sommer was having lunch at the Hamilton Street Club in Baltimore, which is frequented by journalists and literary types. An editor from the Baltimore *Sun* showed him a front-page article from the day before, about Peter Jahrling and his work with smallpox at the CDC, and said, "The USAMRIID people are killing monkeys with smallpox, and they're proud of it. What do you think of that, Al?"

Sommer said that his reaction was, "Excuse me? They're *what*?" He stared at the newspaper and couldn't believe what he was reading. "I started to vibrate at the visceral level," he said. "We could have eradicated smallpox completely if we had just destroyed the stocks a couple of years after the Eradication. And now there was Peter Jahrling exulting in the fact that he could kill these monkeys with smallpox. I went bananas." Sommer was leaving on a trip to Thailand the next morning, but he whipped off an op-ed piece for the paper.

It began: "One needn't be a Luddite to recognize an idiot—and the government scientists gloating . . . over their ability to infect monkeys with smallpox are idiots of the worst sort." Sommer says that the editors wanted to tone him down, so they took out the following sentence: "I am not sure if they are homicidal idiots or suicidal idiots."

He felt that the biggest danger of Jahrling's research was that it would look suspicious to other countries and would encourage them to do their own experimentation. "We could start an arms race over smallpox, and the thinking would go, 'You could be bioengineering smallpox, so I'm going to bioengineer a smallpox, as well.' I don't think it would be hard to bioengineer smallpox," he went on. "My virologist friends are always bioengineering viruses. I could see a bioengineered strain of smallpox getting into a terrorist's hands, and that's my fear. And then when we get a terrorist attack with smallpox, and the smallpox doesn't respond to the vaccine, we're in trouble." He wanted the United States and Russia to get together to destroy their stocks, jointly scour the world for stray stocks of smallpox, and use every effort to persuade other countries to destroy them. He wanted to create an international abhorrence for any nation that would keep smallpox around. He wanted the demon cast out. "It still rankles me," he said, "that we are giving smallpox to animals that could not get smallpox naturally, in order to protect humans, when the last time a human had smallpox was 1978, and humans shouldn't naturally get it today. This is my circular indignation."

I VISITED D. A. Henderson at his home in Baltimore a little over two months after the anthrax attacks. I arrived in the late afternoon, bringing smoked salmon and a bottle of Linkwood malt whisky. Nana Henderson spread out the salmon with lemon and onion on a table in the family room. Their son Doug, who is now a composer, was there. As a teenager, Doug had traveled with his father, and had vaccinated many people himself. In the cool, dry light of a winter's afternoon, the Hendersons and I poured out glasses of Linkwood and picked away at the salmon. D.A. talked about why people had joined the Eradication: "Some of them were looking for themselves, and some of them got involved with feeling what a difference you could make if you could end this disease." The sky began turning to dusk. Pots of dead thyme sat on the deck, silvery and dry. "Smallpox was the only disease we know of for which there were deities," he said. "It was the worst human disease. I don't know of anything else that comes close."

Later, on the subject of Peter Jahrling's work infecting monkeys with variola, Henderson said he was not optimistic that it would lead to new drugs or vaccines. "Do we need to do the research? There are some scientists who feel it's important and should be pursued. But is it really going to work? Peter Jahrling gave the monkeys a huge dose of virus, but it isn't going to be very helpful for testing a new vaccine, because what we really need is an inhaled dose of smallpox in a monkey to test a vaccine, since people inhale the virus." He sounded discouraged, emotionally drained over the fight to destroy the public stocks of smallpox. He was working for the government, and government policy was to look for new cures for smallpox, and that meant doing experiments with variola. He said that he had taken care of his emotions over the issue of destroying the known stocks of smallpox. "Everything is in neutral right now," he said. "There is no point in my entering a battle where the cards are stacked. I'm playing along with what they're doing. I'm asking them to pursue the research." Henderson had gone so far as to suggest to Peter Jahrling that he try an African strain of smallpox, Congo 8, on the monkeys, because it might look more like human smallpox. "If it works, Peter, I want the credit," he said to Jahrling.

"When the research with variola has been pursued to some reasonable point, then I want to revisit the question of destruction," he said. "The subject should be reopened."

He had thanked me for the smoked salmon that day. "It's really large," he remarked. "I wonder: is it one of the newer genetically engineered salmon? It's fairly simple to add one gene to a salmon. Or to any organism in the lab. Will people change organisms in the lab to make them more dangerous? Can it be done? Yeah. Will it be done? Yeah, it will be done," he said. "And there will be unexpected crises."

• • •

ON April 30th, 2002, a group of six experts on the spread of infectious diseases met under conditions of secrecy in a conference room at the John E. Fogarty International Center at the National Institutes of Health (the NIH), in Bethesda, Maryland. Each expert had been asked to create a model of the spread of smallpox in the United States, starting with a small number of infected people. One of the experts—Dr. Martin Meltzer of the CDC—found that smallpox could be easily controlled with ring vaccination using the traditional vaccine. He felt that the virus was not very infective in people and would be unlikely to spread fast or far. The other five experts disagreed with one another, sometimes sharply, but in general they found that smallpox would spread widely and rapidly. They argued forcefully with each other (as scientists do), but in the end, none of the experts could predict what smallpox would do—not to the satisfaction of the other experts. "Our general conclusion was that smallpox is a devastating biological weapon in an unimmunized human population," one of the participants said. "If you look at the real-world data from a 1972 outbreak in Yugoslavia, you find that the multiplier of the virus was ten: the first infected people gave it to ten more people, on average. Basically, if you don't catch the first guy with smallpox before he kisses his wife, it goes out of control. We could be dealing with hundreds of thousands of deaths. It will absolutely shut down international trade, and it will make 9/11 look like a cakewalk. Smallpox can bring the world to its knees." The experts were told by NIH officials that they should not publicize their findings.

SUPERPOX

Dr. Chen's Viruses

UNANSWERED QUESTIONS hung over variola, and not just the question of whether ring vaccination would work if there was a terror attack with smallpox. The more troubling question was how molecular biology would affect the future of smallpox. Poxviruses are used in laboratories all over the world precisely because they are easily engineered. Commercial kits for the process are available at no great cost. It should not be forgotten that the director of the Iraqi virus-weapons program, Dr. Hazem Ali, was a pox virologist trained in England, and one assumes that he is not the only professional bioweaponeer in the world with advanced credentials in biology.

The Australian team of mouse researchers led by Ronald Jackson and Ian Ramshaw had put the IL-4 mouse gene into mousepox and had created a superpox that appeared to break through the mice's immunity. The Jackson-Ramshaw virus was harmless in people, but it seemed to be devastating in immunized mice.

Bioterror planners wondered: if the human IL-4 gene were put into smallpox, would it transform smallpox into a super variola that would devastate immunized humans? The Jackson-Ramshaw virus had been a narrow beam of light shining across a dark landscape of the future. It had shown dim outlines of virus weapons to come.

When an experiment gives a result, the first thing scientists do is try to repeat the experiment to see if they can get the same result. The essence of the scientific method lies in the repeatable result: if you perform an experiment in the same way, nature will do the same thing again. This is the heart of science and is the sign that an observable phenomenon in nature has been found. Would the results of the Jackson-Ramshaw experiment bear out? Could a poxvirus be engineered to crash through a vaccine?

ONE DAY in early 2002, I parked my car in a downtown neighborhood of St. Louis and walked along an uneven sidewalk toward the St. Louis University School of Medicine. The neighborhood is humble but neat, and is largely African-American. There are row houses with porches tucked up against the street. American flags hung from several porches or were on display in windows. The school of medicine is a stately neo-Gothic brick building, trimmed with pink midwestern sandstone, and on that day it glowed with warmth in winter light.

The façade gives way to a concrete, fortresslike structure, five stories tall, with small windows, where the research laboratories are located. In a group of rooms on the fourth floor, a pox virologist named Mark Buller leads a group of researchers who do experiments with mousepox virus and with vaccinia. They work mainly with mice—the mouse is the standard animal used in biomedical research. Most of the important discoveries about how our immune systems work were made originally in experiments done with mice.

Mark Buller is a tall, lanky, self-effacing man in his fifties, a dual citizen of Canada and the United States, with curly black hair, a black mustache, intelligent brown eyes behind round glasses, and a voice that has an attractive Canadian softness. He grew up in Victoria, British Columbia. He often walks around the lab in nylon wind pants, a T-shirt, and running shoes. He keeps a spare jacket and tie hanging on the wall of his office, in case an important meeting comes up. Buller is known and respected among pox virologists, although he seems to deliberately avoid the limelight. "My goal in life is to be prominently in the shadows," he said to me.

Buller began hearing a lot about the Jackson-Ramshaw experiment from Peter Jahrling and Richard Moyer. Right after it was published, Moyer, especially, raised alarms—he began saying, quietly, to Buller that either he or Buller should try to repeat the experiment. The Australian smallpox expert Frank Fenner had advised Jackson and Ramshaw to publish their work, partly on the grounds that nobody would really make an IL-4 smallpox, since it might be too devastating and perhaps even suicidal. In the wake of September 11th, the release of a genetically engineered smallpox into the United States did not seem quite so impossible.

Mark Buller decided to create an IL-4 mousepox, to see if it would blow through a vaccine. He wanted to get a sense of whether a human IL-4 smallpox could become a supervirus, and if so, what vaccination strategy for people would work against it. I arrived at Buller's lab as the experiment was getting under way. I wanted to hold an engineered superpox in my hands and get a feel for where the tide of modern biology was taking us.

MARK BULLER leaned back at his desk, his hands clasped behind his head. His office was crowded with books and papers, and there was an exercise mat on the floor. On a whiteboard on the wall, his daughter, Meghan, had drawn a caricature of him as a science nerd, with Coke-bottle spectacles, a brushy mustache, and a bunch of pens in his shirt pocket.

"If there is a bioterror release of smallpox, currently the main strategy is ring vaccination," he said. "In order for ring vaccination to work, the vaccine has to block severe smallpox disease in people. But what if a smallpox that's expressing IL-4 blocks people's immune responses?"

Buller explained that his group would make four different engineered mousepox virus strains. They would all have the IL-4 gene in them, but they would be slightly different from one another. One of them would be almost the same as the Australian engineered pox. "We want to get a feeling for what the IL-4 gene does in mousepox," Buller said. "I've always found that whenever I try to predict Mother Nature I'm wrong."

Buller's lab was a group of rooms with white floors and cluttered black counters and shelves. Four or five people were working on different projects, and it was a crowded place. In a corner, under a window, a scientist named Nanhai Chen was in the middle of the virus engineering. He was working at a counter that was three feet long and a foot and a half wide. Virus engineering doesn't have to take up much real estate. Mousepox virus, even engineered mousepox, is harmless to humans, because the virus simply can't grow inside the human body, so the work was safe for the people in the room.

Nanhai Chen is a quiet man in his late thirties. He grew up on a collective near Shanghai called the Red Star Farm, where his father was a farmer and where some of his sisters still live. In high school, Chen decided he liked biology, and he went on to have a fast-track career at the Institute of Virology at the Chinese Academy of Preventive Medicine in Beijing, which is probably the top virology center in China. He became an expert in the DNA of vaccinia virus. Mark Buller hired him out of China.

Nanhai Chen has a fuzzy crew cut, hands that work rapidly, wire-rimmed spectacles, and restrained manners. He and his wife, Hongdong Bai, who is also a molecular biologist, have given their children American names, Kevin and Steven. He wears only two outfits, one for winter and one for summer. His winter outfit is a blue cotton sweater, blue slacks, and white running shoes. I spent days with Chen during the time he engineered the mouse supervirus. "It's not difficult to make this virus," he said to me one day. "You could learn how to do it."

A VIRUS that has been engineered in the laboratory is called a recombinant virus. This is because its genetic material—DNA or RNA—has genes in it that come from other forms of life. These foreign genes have been inserted into the virus's genetic material through the process of recombination. The term *construct* is also used to describe it, because the virus is constructed of parts and pieces of genetic code—it is a designer virus, with a particular purpose.

The DNA molecule is shaped like a twisted ladder, and the rungs of the ladder—the nucleotides—can hold vast amounts of information, the code of life. A gene is a short stretch of DNA, typically about a thousand letters long, that holds the recipe for a protein or a group of related proteins. The total assemblage of an organism's genetic code—its full complement of DNA, comprising all its genes—is the organism's genome. Poxviruses have long genomes, at least for viruses. A pox genome typically holds between 150,000 and 200,000 letters of code, in a spaghettilike knot of DNA that is jammed into the dumbbell structure at the center of the pox particle. The poxvirus's genome contains about two hundred genes—that is, the pox particle has around two hundred different proteins. Some of them are locked together in the mulberry structure of the particle. Other proteins are released by the pox particle, and they confuse or undermine the immune system of the host, so that the virus can amplify itself more easily. Poxviruses specialize in releasing signaling proteins that derange control systems in the host. For example, insect poxes release signals that cause an infected caterpillar to stop developing and grow into a bag packed with virus.

The human genome, coiled up in the chromosomes of every typical cell in the human body, consists of about three billion letters of DNA, or perhaps forty thousand active genes. (No one is certain how many active genes human DNA has in it.) The letters in the human genome would fill around ten thousand copies of *Moby-Dick:* a person is more complicated than a pox.

The IL-4 gene holds the recipe for a common immune-system compound called interleukin-4, a cytokine that in the right amounts normally helps a person or a mouse fight off an infection by stimulating the production of antibodies. If the gene for IL-4 is added to a poxvirus, it will cause the virus to make IL-4. It starts signaling the immune system of the host, which becomes confused and starts making more antibodies. But, paradoxically, if too many antibodies are made, another type of immunity goes *down*—cellular immunity. Cellular immunity is provided by numerous kinds of white blood cells. When a person dies of AIDS, it is because a key part of his or her cellular immunity (the population of CD4 cells) has been destroyed by HIV infection. The engineered mousepox seems to create a kind of instant AIDS-like immune suppression in a mouse right at the moment when the mouse needs this type of immunity the most to fight off an exploding pox infection. An engineered smallpox that triggered an AIDS-like immune suppression in people would be no joke.

TO CREATE a construct virus, you start with a cookbook and some standard ingredients. The basic raw ingredient in Chen's experiment was a vial of frozen natural wild-type mousepox virus, which sat in a freezer around the corner from his work area. The other basic ingredient was the mouse IL-4 gene. Chen's cooking, so to speak, involved splicing the gene into the DNA of the poxvirus and then making sure the resulting construct virus worked as it was supposed to.

Chen ordered the IL-4 gene through the Internet. It cost sixty-five dollars, and it came by regular mail at Mark Buller's lab in November 2001, from the American Type Culture Collection, a nonprofit institute in Manassas, Virginia, where strains of micro-organisms and common genes are kept in archives. The gene arrived in a small, brown glass bottle with a screw top. Inside the bottle was a pinch of tan-colored dry bacteria—*E. coli*, bacteria that live in the human gut. The bacterial cells contained small rings of extra DNA called plasmids, and the plasmids held the IL-4 gene. The IL-4 gene is a short piece of DNA, only about four hundred letters long, and it is one of the most common genes used in medical research. To date, more than sixteen thousand scientific papers have been written on the IL-4 gene.

The standard cookbook for virus engineering is a four-volume series in ring binders with bright red covers, entitled *Current Protocols in Molecular Biology,* published by John Wiley and Sons. Nanhai Chen took me to a shelf in the lab, pulled down volume three of *Current Protocols,* and opened it to section 4, protocol 16.15, which describes exactly how to put a gene into a poxvirus. If anyone puts the IL-4 gene into smallpox, they may well do it by the book. "This cannot be classified," Chen said, running his finger over the recipe. "No one ever thought this could be used for making a weapon. The only difficult part of it is getting the smallpox. If somebody has smallpox, all the rest of the information for engineering it is public."

"Are you personally worried about engineered smallpox?"

"Yes, I am," he answered, holding the cookbook open as he spoke. "I was talking last week with my mentor in China. His name is Dr. Hou, and he's a very famous virologist in China. He told me the Russians have a genetically modified and weaponized smallpox. My mentor didn't say where he learned this, but I think he has good access to information, and I think it is probably true. Smallpox was all over the world thirty years ago. It could be anywhere today. It's not hard to keep back a little bit of smallpox in a freezer."

I will omit the subtleties of Chen's work for the sake of general readers, but the outline of a recipe for making the biological equivalent of an atomic bomb is in these pages. I would hesitate to publish it, except that it's already known to biologists; it just isn't known to everyone else. It doesn't take a rocket scientist to make a superpox. You do need training, though, and there is a subtle art to virus engineering. One becomes better at it with experience. Virus engineering takes skill with the hands, and in time you develop speed. Chen felt that with a little luck he could engineer any sort of typical construct poxvirus in about four weeks.

Chen took the little brown glass bottle of dry bacteria that contained the IL-4 gene and cultured the bacteria in vials. Then he added a detergent that broke up the bacteria, and he spun the material in a centrifuge. The cell debris fell to the bottom of the tubes, but the DNA plasmid rings remained suspended and floating in the liquid. He ran this liquid through a tiny filter. The filter trapped the DNA that held the IL-4 gene. He ended up with a few drops of clear liquid.

Next, Chen spliced some short bits of DNA, known as promoters and flanking sequences into the plasmid rings. He did this basically by adding drops of liquid. Promoters signal a gene to begin making protein. The various promoters were going to cause the strains of engineered mousepox to express the IL-4 protein in differing amounts and at different times in the life cycle of the virus as it replicated in cells.

The next step was to put the engineered DNA into the virus, using a genetic-engineering kit called a transfection kit. Transfection is the introduction of foreign DNA into living cells. A transfection kit is essentially a small bottle filled with a reagent, or biochemical mix; a bottle of it costs less than two hundred dollars. You can order transfection kits in the mail from a variety of companies. Nanhai Chen used the Lipofectamine 2000 kit from Invitrogen.

Chen grew monkey cells in a well plate, and then he infected them with natural mousepox virus. He waited an hour, giving the virus time to attach to the cells. Then he added the IL-4 DNA, which he'd already mixed with the transfection reagent. He waited six hours. During that time, the IL-4 DNA was taken up into the monkey cells, which were also infected with natural mousepox. Somehow, the IL-4 DNA went into some of the mousepox particles, and the IL-4 gene ended up sitting in the DNA of the mousepox virus.

Chen had long days of work ahead of him, for he had to purify the virus strains. Purification of a virus is a core technique in the art of virus engineering.

• • •

A VIRUS is a very small object, and the only way to handle it is to move around cells that are infected with it. A poxvirus growing in the layer of cells at the bottom of a well plate will kill the cells, forming dead spots in the layer. These spots are like the holes in a slice of Swiss cheese, and they are known as plaques. You can remove the dead or dying cells with a pipette. The cells that come out of that spot will contain a pure strain of the virus.

"Would you like to do some plaque picking?" Chen asked me one day. He led me into a small room behind his work area, where there were a couple of laboratory hoods, a couple of incubators (which are warming boxes that keep cell cultures alive), and, tucked away in a corner, a microscope with binocular eyepieces.

Chen put on a pair of latex gloves, opened the door of an incubator, and slid out a well plate. It had six wells, glistening with red cell-culture medium, and a carpet of living cells covered the bottom. He carried the well plate across the room and placed it on the viewing stand of the microscope. You could see with the naked eye the holes in the cell layers. The cells were infected with a strain of engineered IL-4 mousepox.

I sat down at the microscope, and Chen handed me a pipette that had a cone-shaped plastic tip with a hole in it, like a very fine straw. You put your thumb on a button on the pipette, and when you pushed the button you could pick up a small amount of liquid and deposit it somewhere else.

I was beginning to feel a little strange. We were handling a genetically engineered virus with nothing but rubber gloves. "You're sure it's not infective?"

"Yes, it is safe."

I sat down at the microscope and looked into a carpet of monkey cells growing at the bottom of a well. Each cell looked like a fried egg; the yolk in the cell was the nucleus. I started looking for holes in the carpet, where the virus would be growing.

"I can't find any plaques," I said. I began moving the well plate around. Suddenly, a huge hole appeared. It was an infected zone, rich with engineered virus. The cells there were dying and had clumped up into sick-looking balls. The cells had caught the engineered pox.

I was holding the pipette in my right hand. I maneuvered the tip into the well plate. "I can't see the tip," I said, jabbing it around in the well.

I was wrecking Chen's careful work, but he made no comment. Then the tip of the pipette heaved into view. It looked like the mouth of a subway tunnel.

"You need to scratch the cells off," Chen said.

I moved the tip around, scraping it over the sick cells. I let the button go, and a few cells were slurped up into the pipette. Chen handed me a vial, and I deposited a picked plaque of engineered poxvirus into it. "I don't think I'd make a good virologist."

"You are doing fine."

The work of creating four engineered mousepox strains took five months—the work was painstaking, and Chen had to check and double-check every step of the process. He believes that the total cost of laboratory consumables ran to about a thousand dollars for each strain. Virus engineering is cheaper than a used car, yet it may provide a nation with a weapon as intimidating as a nuclear bomb.

IT WAS TIME to infect some mice with the engineered virus, to see what it would do. The mouse colony was kept in a Biosafety Level 3 room on the top floor of the medical school. Mark Buller and I put on surgical gowns, booties, hair coverings, and latex gloves. We pushed through a steel door into a small cinder-block room, where hundreds of mice were living in clear plastic boxes, set on racks behind glass doors. The mice had black fur. They were a purebred laboratory mouse known as the Black 6, which is naturally resistant to mousepox.

Buller opened some boxes, removed some mice, and placed them in a jar that had an anesthetic in it. The mice went to sleep. One at a time, he held a mouse in his hand, stuck the needle of a syringe into its foot, and injected a drop of clear liquid. The liquid contained about ten particles of engineered IL-4 mousepox—an exceedingly low dose of the virus.

Seven days later, my phone rang early in the morning. It was Mark Buller. One of the lab techs had just checked on the mice, he said, and some of them had a hunched posture, with ruffled fur at the neck. "They're going to go fast," he said.

The next morning, Buller, Chen, and I put on gloves and gowns and went into the mouse room. There were two boxes of dead mice. Two of the strains of IL-4 mousepox had wiped out the naturally resistant mice. The death rate for those groups was one hundred percent.

Buller carried one box inside a hood and opened it. The dead mice were indeed hunched up, with ruffled fur and pinched eyes. Natural mousepox does not cause a Black 6 mouse to become visibly sick at all.

"Wow. Wow," Chen said. "They're all hunched over. This IL-4 has a really funny effect. This is really a strong virus. I'm really surprised." He hadn't expected his virus to wipe out *all* the mice. It disturbed him that he could make such a powerful virus, but he also felt excited.

"It's really impressive how fast this virus kills the mice at such a low dose," Buller said.

I sat on a chair before the hood, peering into it beside Buller. He reached in and lifted a dead mouse out of a box, and held the creature in his gloved hand. Without the mouse, there would be no cures for many diseases, and dead mice had been responsible for the saving of many a human life, but what he held in his hand was not a reassuring thing.

Buller showed me the standard way to dissect a mouse: you slit the belly with scissors. He spread open the abdomen with the scissors, looking to see what the pox had done.

The virus had blasted the mouse's internal organs. The spleen had turned into a bloated blood sausage that was huge (for a mouse's spleen) and filled much of the mouse's belly. It was mottled with faint grayish-white spots, which Buller explained is the classic appearance of a mouse's organs infected with pox. Doctors who opened humans who had died of hemorrhagic smallpox saw the same cloudy effect in their organs. With the tip of the scissors, he pulled out the mouse's liver. It had turned the color of sawdust, destroyed by the engineered virus. With ten particles of the construct virus in its blood, the pox-resistant mouse had never stood a chance.

THERE ARE TWO WAYS to vaccinate a mouse against mouse-pox. One way is to infect it with natural mousepox. When it recovers (if you vaccinate a resistant breed of mouse, it will recover), it will be immune. The other way is to vaccinate the mouse with the smallpox vaccine—that is, you infect the mouse with vaccinia, and its immunity to mousepox goes up in the same way that a human's resistance to smallpox goes up after a vaccinia infection.

Mark Buller and his group began testing IL-4 mousepox on vaccinated mice, and they got strange results. They were not able to completely duplicate the Jackson-Ramshaw experiment. They discovered that mice immunized with natural mousepox become completely immune to IL-4 mousepox—it did *not* break through their immunity after all. That was very encouraging. It contradicted part of the Jackson-Ramshaw experiment. But in doing preliminary experiments with the smallpox vaccine, they had begun to see something more troubling (the experiments were in progress, and Buller wasn't able to report any real findings yet). It seemed that the IL-4 mousepox could crash through the smallpox vaccine, killing the mice if they had been vaccinated sometime previously. But if their vaccinia vaccinations were very fresh, they were protected against the engineered pox. It suggested that an engineered IL-4 smallpox might be able to break through people's immunity, but not if the vaccinations were recent, perhaps only weeks old.

Buller didn't sound as if he thought the world was coming to an end. "We showed that you *could* find a way to vaccinate mice successfully against the engineered mousepox," he said to me. "Even if IL-4 variola can blow through the smallpox vaccine, I feel there are drugs we can develop that will nullify the advantage a terrorist might have by using IL-4 variola. We really need an antiviral drug," he said. He argued that a drug that worked on pox was not only needed as a defense against an engineered superpox, but was also needed in order to cure people who were getting sick from the vaccine during a mass vaccination after a smallpox terror attack.

Any nation or research team that wanted to make a superpox would have to test it on vaccinated humans to see if it worked. "If you're talking about a country like Iraq," Buller said, "human experimentation with smallpox is imaginable. If you've got a guy like Saddam Hussein, and his scientists tell him they need some humans so they can check out an engineered smallpox, he'll say, 'How many do you need?' There are people like that in every age."

Nanhai Chen seemed a little less optimistic. "Because

the IL-4 mousepox can evade the vaccinia vaccination, it means that IL-4 smallpox could be very dangerous," he said. "This experiment is very similar to the human situation with the smallpox vaccine. I think IL-4 smallpox is dangerous. I think it is very dangerous."

THE MAIN THING that stands between the human species and the creation of a supervirus is a sense of responsibility among individual biologists. Given human nature and the record of history, it seems possible that someone could be playing with the genes of smallpox right now. And what if a fire began to flicker in the hay in the barn, and we poured a glass of water on it, but the water could not put the fire out? No nation that wanted to have nuclear weapons had a problem finding physicists willing to make them. The international community of physicists came of age in a burst of light over the sands of Trinity in New Mexico. The biologists have not yet experienced their Trinity.

A Child

IN THE YEARS just before the Eradication began, two million people a year were dying of smallpox. The doctors who ended the virus as a natural disease have effectively saved fifty to sixty million human lives. This is the summit of Everest in the history of medicine, and yet they have never received the Nobel Prize. These days, several times a year, Dr. Stanley O. Foster takes a trip to revisit places he's worked at before. In 2000, he decided to take a cruise aboard the *Rocket*. He arrived at the landing at Dhaka, carrying a small knapsack, and there she was at the dock, the steam paddle wheeler, completely unchanged, looking exactly as she had in 1975, stained with rust and jammed with humanity. He spent the night aboard her, leaning on the rail and watching the islands pass, smelling the river and the rising scent of the sea, and shortly after sunrise he disembarked at Berisal, where he rented a speedboat and crossed the bay to Bhola Island. He went inland by Land Rover, threading his way among crowds of people, until he came to the house of Rahima.

The young woman had moved to a different village when she had married. She was now twenty-five years old, and she was most happy to see him, though she was just about as shy as she had been on the day when she had dived into the burlap sack. Rahima had two daughters, was expecting a third child, and was hoping for a son. She presented Dr. Foster with a small gift, and he gave her kids some crayons.

AT SUNRISE one day in November 2001, a month after the anthrax attacks, I drove south through Gettysburg, past the Gettysburg battlefield. It is an open country of rolling farms, and it looks not very different than it did at the time of the Civil War. The earth was a rich brown, dotted with crows, which flew up into a yellow sky. Little Round Top passed, the hill where Joshua Chamberlain and his men from Maine turned the tide with a charge. It was just another hump of bare trees. The road took me to Frederick, and I walked through the corridors of USAMRIID, along with Peter Jahrling, Lisa Hensley, and Mark Martinez. The light was sickly green, and the air smelled of roasting equipment—giant autoclaves baking things to sterilize them. Corridors branched to the left and right. Martinez was dressed in fatigues, with a black beret tucked in his belt. He swiped his security card over a sensor and pushed open a door, and we walked into the pathology suite—a cold suite, where there are no dangerous pathogens. Martinez took us into a small, windowless room with green walls. It was a bare-bones room, with some filing cabinets, some work counters, and a hood.

"I'll be right back," Martinez said and left the room.

We leaned up against the cabinets and waited. He was going to a storage area that was, apparently, a secret.

"What you are about to see is a national resource," Jahrling remarked.

"I've never seen it," Hensley commented.

Martinez returned, carrying a white plastic bucket. He popped open the lid and removed something that was wrapped in a yellow disposable surgical gown inside a plastic bag. He placed the bag inside the hood, opened it, and slid out the lump wrapped in the gown. Very slowly and carefully, he peeled the gown away and revealed it.

It was the arm of a child, covered with smallpox pustules. The arm had been severed during autopsy.

The child had been American, white, three to four years old, and had died of variola major. That was about all the information the Army scientists had been able to come up with.

In the spring of 1999, a professor at the Indiana University School of Dentistry had been exploring a dark basement corridor of the school with a flashlight, and he had come across a collection of jars that had belonged to

William Schaffer, a long-dead pathologist. One of them was a jar with the arm in it, labeled "M 243 Smallpox." No one knew where Professor Schaffer had obtained it. The professor had phoned officials at the CDC and had asked them if they wanted it. The CDC did not want a pickled arm covered with smallpox, so the professor gave it to a pharmaceutical company that was working on drugs for smallpox. Jahrling had shown up at the company one day to talk about smallpox drugs, and company scientists had mentioned to him that they had a smallpox arm, and did he want it?

"Heck, yes," Jahrling said to them.

There weren't any living people with the disease, and an arm covered with variola pustules was a magnificent clinical specimen. He had wanted to wrap the arm in plastic and bring it back in his carry-on luggage, but he began to wonder what airport security people would do if they found it, so he made arrangements to have the arm shipped to USAMRIID by express delivery.

The WHO forbids any laboratory except the CDC and Vector to have more than ten percent of the DNA of the smallpox virus. The chemicals in the jar had caused the smallpox DNA to fall apart into tiny fragments, and thus it was a legal smallpox arm.

THE ARM was lying palm downward. Mark Martinez turned it slowly, holding it with great care, until the palm came upward. He took the index finger and bent it ever so slightly, opening the palm, revealing the erupting centrifugal rash. The arm was covered with dark brown pustules. The child had died at the moment the pustules were beginning to crust over. The crusts were very dark. Lisa Hensley stared at it through the glass. "I never had any idea how bad smallpox was until I saw the lesions in the monkeys. You can see the same lesions here."

Martinez stood up, leaving the arm inside the hood. I put on a pair of latex gloves, sat down on the stool, reached into the cabinet, and picked up the arm. I could smell a faint, sweet smell of preserved human flesh. For a moment, I wondered if it was the foetor of smallpox. I turned the arm over and scabs began falling onto my hands. The lifeboats of variola were coming off.

THE APPEARANCE of a child's arm covered with small-pox pustules was something that our ancestors knew, but the arm had become a relic of history and an object of horror, alien to us. That we had never seen an arm like this in our lives was an extraordinary thing, a gift, unasked for, unexpected, and now unnoticed. A handful of doctors had given it to us, joined by thousands of village health workers. They had forged themselves into an army of peace. With a weapon in their hands, a needle with two points, they had searched the corners of the earth for the virus, opening every door and lifting every scrap of cloth. They would not rest, they would not stand aside, and they gave all they had until variola was gone. No greater deed was ever done in medicine, and no better thing ever came from the human spirit.

As I reflected on the death of variola, I thought also about our future. More and more people are living in cities. Soon more than half of the people in the world will be city dwellers. According to projections made by the United Nations, by the year 2015, the earth will likely contain twenty-six extremely big cities. Twenty-two of those cities will be in developing nations. Perhaps four of them will be in industrial countries. New York and Los Angeles will be medium-sized cities then.

These big cities will be in tropical nations. Bombay will have twenty-six million people living in it by 2015, Lagos twenty-four million. The population of California is currently thirty-five million. Take two thirds of the people in California and cram them into one city, with poor sanitation and inadequate health care and ineffective government. Twenty-five million people living within a couple of hours of one another . . . this is a leap beyond any sort of crowding a poxvirus would have found in ancient Egypt. If there is not enough vaccine to stop an outbreak of smallpox in a giant city, or if the virus cuts through the vaccine because human beings have done something to its genes, then the virus will move fast. The cities of the world are linked through a web of airline routes. A virus that appears in Bangladesh will soon arrive in Beverly Hills. An engineered virus could bring a bit of invisible bad news to every community on earth.

I OPENED the child's hand and spread the fingers out on my palm. The child's palm was one entire pustule. The fingers had gone confluent, so that there was almost no skin left that was not pustulated. I could see the whorls of the fingerprints, the mounts of fate and the future. The line of life and the line of love had been broken.

What was not present any longer in this hand was the suffering of the child who had endured it. I recalled when my son had been born, and how, minutes afterward, I had held his tiny hand, impressed with its perfection. I recalled the times when my children had been sick or had needed comforting, and I had held their hands. I recalled the sense of time slipping by as I noted how their hands were growing, gradually filling more of my hand. They might one day hold my hand when the life had left it.

We will never find an explanation for the suffering etched in that child's hand, or for the evils done by people against other people, or for the love that drove the doctors to bring smallpox to an end. Yet after all they had done, we still held smallpox in our hands, with a grip of death that would never let it go. All I knew was that the dream of total Eradication had failed. The virus's last strategy for survival was to bewitch its host and become a source of power. We could eradicate smallpox from nature, but we could not uproot the virus from the human heart.

Glossary

amplification. Multiplication of a virus. See **replication.**

anthrax. *Bacillus anthracis,* a rod-shaped, spore-forming bacterium that grows profusely in lymph and blood. Name comes from the Greek word for *black,* after the blackening of the skin caused by an anthrax infection of the skin.

anthrax spores. Tiny ovoid spores, one micron long, produced when anthrax bacterial cells encounter adverse conditions and are unable to keep growing. About two hundred spores would span the thickness of a human hair.

antiviral drug. A type of drug that stops or slows a virus infection.

biological weapon or **bioweapon.** Disease-causing pathogen dispersed into a human population as a weapon. Usually prepared and treated in special ways in order to be dispersed in the air.

Biosafety Level 4. Also **BL-4** or **Level 4.** Highest level of biocontainment; requires the wearing of a bioprotective space suit.

black pox. See **flat hemorrhagic smallpox.**

blue suit. A bioprotective full-body space suit, typically blue.

BWC. The Biological Weapons and Toxin Convention,

an international treaty, signed by more than one hundred and forty nations, which forbids the development, possession, and use of offensive biological weapons. Signed by the United States in 1972.

CDC. The federal Centers for Disease Control and Prevention, in Atlanta.

chains of transmission. Chains of infection, which typically branch through a population.

construct. A recombinant virus made in the laboratory.

cytokine. A signaling compound, released by cells, that circulates in the blood and lymph and regulates a system in the body. Many cytokines serve as signals in the immune system.

cytokine storm. Derangement and collapse of the immune system and other systems in the body.

DNA. The long, twisted, ladderlike molecule that contains the genetic code of an organism. The rungs of the ladder, or nucleotide bases, are the letters of the code.

dumbbell core. Dumbbell-shaped body in the center of a **poxvirus** particle, which contains the virus's **DNA** or genome. Also known as the dogbone of pox. Unique to poxviruses.

engineered virus. A recombinant virus or **construct** virus that contains foreign genes in its **DNA** or RNA.

epidemiology. The science and art of tracing the origin and spread of diseases in populations with the goal of controlling or stopping them.

flat hemorrhagic smallpox. Also known as black pox. The most malignant form of smallpox disease, characterized by hemorrhages and darkening (purpuric) skin. It is essentially one hundred percent fatal.

gene. A short stretch of **DNA** that contains the genetic code for a single protein or a related group of proteins in an organism.

genetic engineering. The science and art of inserting

genes into or removing them from the **DNA** of an organism, changing the organism's inherited characteristics as a life-form.

genome. The entire amount of **DNA** in a cell or virus particle, which contains the complete genetic code of the organism. (Some viruses use RNA for their genomes.)

Harper strain. Hot strain of smallpox in the smallpox repository at the **CDC**.

HMRU. The Hazardous Materials Response Unit of the FBI. A rapid-response team for incidents of chemical or biological terrorism; stationed at the FBI Academy in Quantico, Virginia.

host. An organism that a **parasite** lives inside or on.

hot. Virulent; infective; lethal.

IL-4 gene. The gene that codes for interleukin-4, which is a common **cytokine** that regulates the immune system.

IL-4 smallpox. A genetically engineered smallpox not known to exist, though some experts fear it might easily be created in a laboratory by the insertion of the human **IL-4 gene** into natural smallpox virus. Recent experiments suggest that IL-4 smallpox might evade the vaccine and be superlethal in humans. See **mousepox** and **IL-4 gene.**

India-1 strain. A strain of smallpox, believed to be exceedingly virulent in humans, possibly vaccine resistant, weaponized, and produced in tonnage quanties by the Soviet Union for loading into ICBMs.

laminar-flow hood or simply **hood.** A laboratory cabinet with a sliding glass front, similar in principle to an exhaust hood over a kitchen stove, used to protect samples from becoming contaminated and researchers from becoming infected.

micron. One millionth of a meter. An **anthrax spore** is one micron long. Bioweapons particles are ideally one to five microns in size, so as to be inhaled deeply into the lungs.

mirrored smallpox. A doubled collection of smallpox, kept in identical, or "mirror," freezers designated A and B. If one freezer is lost, the smallpox collection remains intact in the other freezer.

mousepox virus. Also called ectromelia. A poxvirus of mice that is related to smallpox. IL-4 mousepox is a genetically engineered mousepox that breaks through some vaccine-induced immunity in some types of mice. See **IL-4 smallpox.**

multiplier. An estimate of the number of secondary cases that will be caused by one infectious case. Known technically as R-zero.

nanopowder silica. Extremely fine particles of silica glass, which can be mixed into a **biological weapon** to make it better able to become easily airborne and thus more infective in the lungs.

parasite. An organism that lives inside or on a **host** organism and typically harms the host.

pipette. Handheld push-button device used for moving small amounts of liquid.

plaque picking. Method of using a **pipette** to suck cells infected with a virus out of a **well plate.** Technique for purifying a strain of **engineered virus.**

plasmid. A short piece of **DNA,** in the shape of a ring, that multiplies inside bacteria as they grow. A plasmid can be engineered with foreign genes and then recombined with a virus to make an **engineered virus.**

poxvirus. A large family of viruses, found in mammals, reptiles, amphibians, birds, and insects. Poxvirus particles are among the largest and most complex virus particles in nature.

Rahima strain. A strain of smallpox in the repository at the **CDC,** taken from scabs of Rahima Banu, a three-year-old girl in Bangladesh who was the last person on earth to be naturally infected with variola major.

replication. Self-copying. See **amplification.**

ring vaccination. Prophylactic technique of vaccinating every susceptible person within a ring around an outbreak.

trans-species jump. The process whereby a virus changes types of hosts, moving from one species to another.

USAMRIID. United States Army Medical Research Institute of Infectious Diseases, at Fort Detrick, Maryland. Also known as the Institute or Rid.

vaccine. A compound or virus that, when introduced into the body, provokes immunity to a disease.

vaccine breakthrough. A (typically lethal) infection that breaks through a person's vaccine-induced immunity.

vaccinia. A **poxvirus** closely related to smallpox. It is much less virulent than smallpox in humans and is used as the vaccine for it.

variola. Scientific name for smallpox virus; comes in two natural subtypes, variola major and variola minor.

Vector. The State Research Center of Virology and Biotechnology, near Novosibirsk, Siberia.

virion. **Virus** particle.

virulence. Ability to cause disease; lethality.

virus. The smallest form of life, a parasite that can replicate only inside cells, using the cell's machinery. Viruses are small particles made of proteins, with a core containing **DNA** or RNA.

virus weapon. A virus that has been prepared for use as a weapon. May be made through **genetic engineering.**

well plate. A plastic plate divided into cups or wells, where viruses are grown inside living cells; essential tool for virus engineering.

World Health Organization (WHO). International body associated with the United Nations; headquartered in Geneva.

Acknowledgments

I AM most grateful for the enthusiastic support and guidance of many people at Random House: Ann Godoff, Joy de Menil, Carol Schneider, Liz Fogarty, Daniel Rembert, Carole Lowenstein, Sybil Pincus, Laura Wilson, Allison Heilborn, Timothy Mennel, Robin Rolewicz, Evan Camfield, Lynn Anderson, Laura Goldin, Richard Elman, Ivan Held, and Laura Wilson (Random House Audio Books).

At Janklow & Nesbit Associates, Lynn Nesbit, Cullen Stanley, Tina Bennett, Bennett Ashley, Amy Howell, and Kyrra Rowley have been incredibly supportive and effective.

The Alfred P. Sloan Foundation and its grant officer Doron Weber provided a highly useful research grant.

Oliver Eickhoff, of the *Westfalenpost*, in Meschede, Germany, helped me research the 1970 Meschede outbreak. Prof. Werner Slenczka gave guidance, Magdalena Drinhaus (Geise) shared her recollections, and Dr. Beate Smith translated documents and interviews.

Andy Young provided valuable professional fact checking.

Numerous people gave important assistance, and shared their thoughts, during the research and writing.

Not all are mentioned in the text, but all contributed: Ken Alibek, Charles Bailey, Daria Baldovin-Jahrling, Dr. Ruth Berkelman, Dr. Michael Bray, Dr. Joel Breman, Dr. Larry Brilliant, Dr. Mitchell Cohen, Richard J. Danzig, Dr. Christopher J. Davis OBE, Louise Davis, Annabelle Duncan, Joseph Esposito, Dr. David Fleming, Dr. Stanley O. Foster, Dr. Mary Frederick, Tom Geisbert, Celia (Sands) Hatfield, Doug Henderson, Dr. D. A. Henderson, Leigh Henderson, Nana Henderson, Dr. Michael Hensley, Lisa Hensley, John W. Huggins, Dr. James M. Hughes, Martin Hugh-Jones, Dr. Thomas Inglesby, Peter B. Jahrling, Dr. Laura Kahn, David Kelly, Dr. Jeffrey Koplan, Dr. Thomas G. Ksiazek, Dr. James LeDuc, Dr. Frank J. Malinoski, Mark J. Martinez, Richard W. Moyer, Randall S. Murch, Frederick A. Murphy, Margaret Nakano, Dr. Tara O'Toole, Michael Osterholm, Dr. John S. Parker, Virginia Patrick, William C. Patrick III, Dr. Brad Perkins, Dr. C. J. Peters, Edward M. Phillips, Tanja Popovic, Rosemary Ramsey, Drew C. Richardson, Dr. Philip K. Russell, Dr. Alfred Sommer, Richard Spertzel, Lisa Swenarski, Shirley Tilghman, Wavy Gravy, Dr. Paul F. Wehrle, John Wickett, Tom Wilbur, Shieh Wun-Ju, Dr. Sherif Zaki, Dr. Alan Zelicoff.

At the FBI: R. Scott Decker, Arthur Eberhart, Philip Edney, Peter Christopher Murray, Scott Stanley, Rex Tomb, and David Wilson.

At St. Louis University School of Medicine: Hongdong Bai, Cliff Bellone, Mark Buller, Nanhai Chen, David Esteban, Joe Muehlenkamp, Gelita Owens, and Jill Schriewer.

Sharon DeLano, who served as my editor for *The Hot Zone* and *The Cobra Event*, also edited this book. She has therefore been the editor of what I think of as a trilogy on Dark Biology.